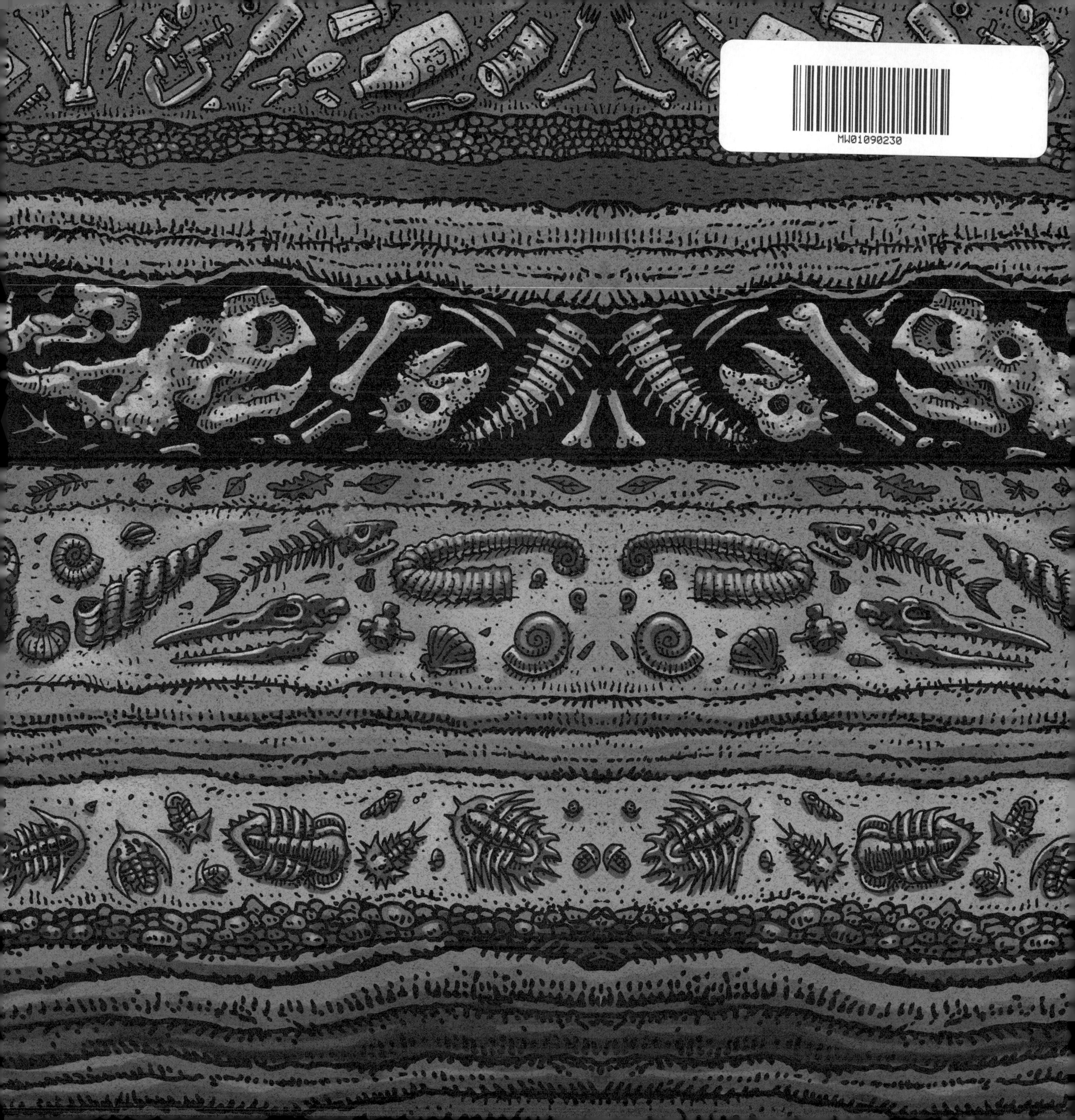
MW01090230

FOR TIM - ROCK ON!
CRUISIN' THE FOSSIL COASTLINE
THE TRAVELS OF AN ARTIST AND A SCIENTIST ALONG THE SHORES OF THE PREHISTORIC PACIFIC
KIRK JOHNSON (THE SCIENTIST)
RAY TROLL (THE ARTIST)

Library of Congress Cataloging-in-Publication Data

Names: Johnson, Kirk R., 1960- author. | Troll, Ray, 1954- illustrator.
Title: Cruisin' the fossil coastline : the travels of an artist and a scientist along the shores of the prehistoric pacific / written by Kirk Johnson ; illustrated by Ray Troll.
Description: Golden, CO : Fulcrum Publishing, [2018] | Audience: Age 12+ |
Audience: Grade 7 to 8. | Includes index.
Identifiers: LCCN 2018013103 | ISBN 9781555917432
Subjects: LCSH: Fossils--Pacific Coast (North America)--Juvenile literature.
Classification: LCC QE714.5 .J634 2018 | DDC 560.979--dc23
LC record available at https://lccn.loc.gov/2018013103

Project Manager: Rebecca McEwen
Editor: Alison Auch
Digital color by: Grace Freeman: IV-V, VII, X, 5, 6, 9, 21, 33, 36-37, 43, 49, 56-57, 76-77, 83, 84, 90, 96, 101, 109, 123, 142, 147, 148, 161, 162-163, 170-171, 182-183, 184-185, 209, 210-211, 212-213, 214, 218, 231 (bottom), 232-233, 245, 246; Ray Troll and Memo Juaregui collaborative acrylic mural VI-VII; Memo Juaregui: XII-1, 2-3, 118-119; Terry Pyles: 116-117

Printed in South Korea
0 9 8 7 6 5 4 3 2 1

Fulcrum Publishing
4690 Table Mountain Drive, Suite 100
Golden, CO 80403
800-992-2908 • 303-277-1623
fulcrum.bookstore.ipgbook.com

**FOR CHASE, PAST AND FUTURE.**
**—K. J.**

**FOR MY "IRISH TWIN" SISTER MIMI AND FOR SEAN DURAN WHOSE FRIENDSHIP AND WISDOM TOUCHED SO MANY.**
**—R. T.**

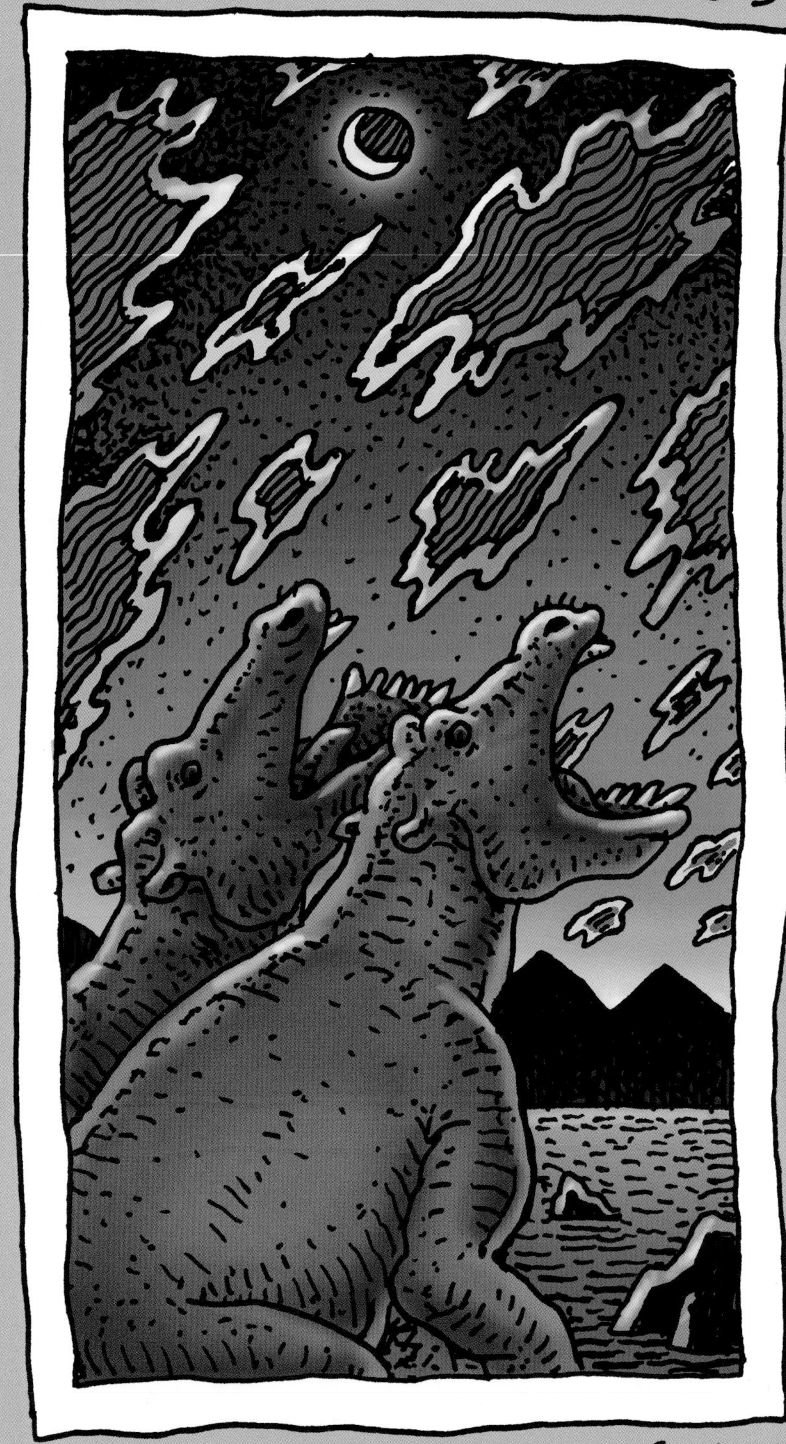
BELLOWING DESMOS
IN THE TWILIGHT

# TABLE OF CONTENTS

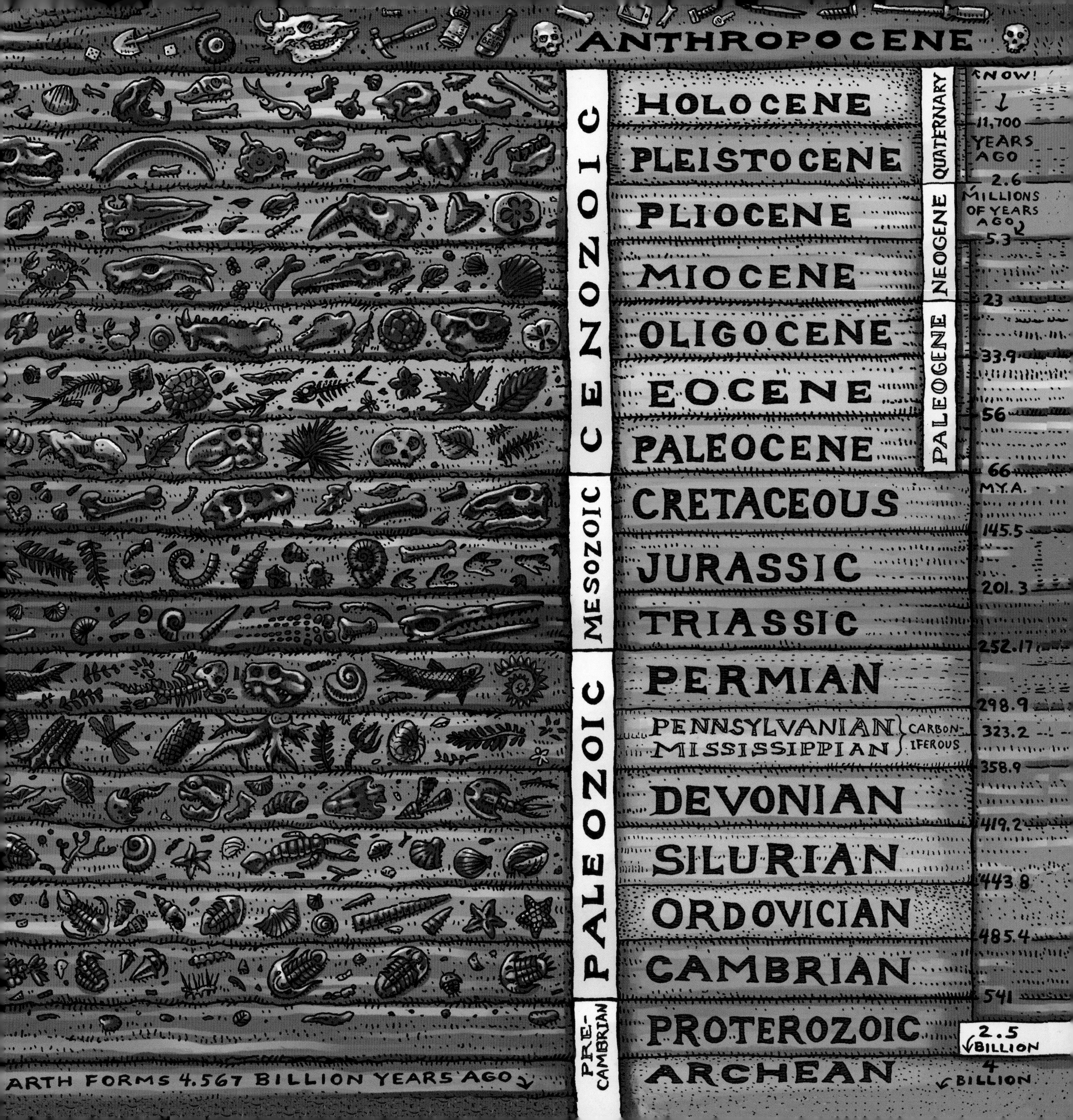

ANTHROPOCENE

CENOZOIC

QUATERNARY
HOLOCENE
NOW!
11,700 YEARS AGO
PLEISTOCENE
2.6 MILLIONS OF YEARS AGO

NEOGENE
PLIOCENE
5.3
MIOCENE
23

PALEOGENE
OLIGOCENE
33.9
EOCENE
56
PALEOCENE
66 M.Y.A.

MESOZOIC
CRETACEOUS
145.5
JURASSIC
201.3
TRIASSIC
252.17

PALEOZOIC
PERMIAN
298.9
PENNSYLVANIAN } CARBONIFEROUS
323.2
MISSISSIPPIAN
358.9
DEVONIAN
419.2
SILURIAN
443.8
ORDOVICIAN
485.4
CAMBRIAN
541

PRE-CAMBRIAN
PROTEROZOIC
2.5 BILLION
ARCHEAN
4 BILLION

ARTH FORMS 4.567 BILLION YEARS AGO

# PROLOGUE

The curious thing about the West Coast of North America is that it has been a coast for a very long time. That statement seems obvious, and perhaps not even worth stating, but it turns out to be true, even when viewed from the inky depths of geologic time. This coastline has been around for more than 750 million years.

On Earth, coastlines come and coastlines go. They do this through the slow but steady process of plate tectonics. Over the long history of the planet, continents split and pull apart, forming half-continents that face each other and then slowly recede from each other's view. The Atlantic Ocean was formed by this process. Two hundred million years ago, the Atlantic wasn't even wet – it was the place where North and South America were stuck to Europe and Africa. But once the split occurred, a narrow sea formed, eventually widening to become an ocean that is still getting wider today. And when the Atlantic formed, it had two coasts that had not been there before.

*Helicoprion*, also known as the buzz saw or whorl-toothed shark, lived during the Permian Period and roamed the shores of western North America.

This drags the edge of the continent down into the Earth where it heats up and eventually melts, forming liquid rock known as magma. Some of this liquid rock, full of gas like a carbonated beverage, squirts up through cracks and erupts at the surface, forming volcanoes and lava flows.

Coastal mountains also form when chains of ocean islands are carried into slow motion collisions with continents. Rather than sliding under the continents, the island chains become welded to them, adding land and growing the continent at the expense of the ocean. This means that the Pacific Coast has been growing to the West, and older versions of it can be found in places such as Nevada and Idaho. Like many of us, North America has grown wider as it has grown older.

It is not just the edge of the continent that is being modified, though. Sea level rises when melting ice sheets make more seawater or when the eruption of seafloor volcanoes decrease the volume of the ocean basin. It falls when ice sheets form or when seafloor volcanic activity

As for those coastlines that go away? Consider this: Before India collided with Asia about 40 million years ago, it had a north coast and Asia had a south coast. Those coasts were then smashed together and thrust into the sky, forming the Himalayas, the world's highest mountain range. Two coasts became one mountain range.

Mountains are often found along coasts, and this is no coincidence. Mountains are formed when the rock at the deep bottom of an ocean is pushed into a continent. The deep ocean rock is more dense than the rocks of the continent, so the ocean floor is driven below the continent.

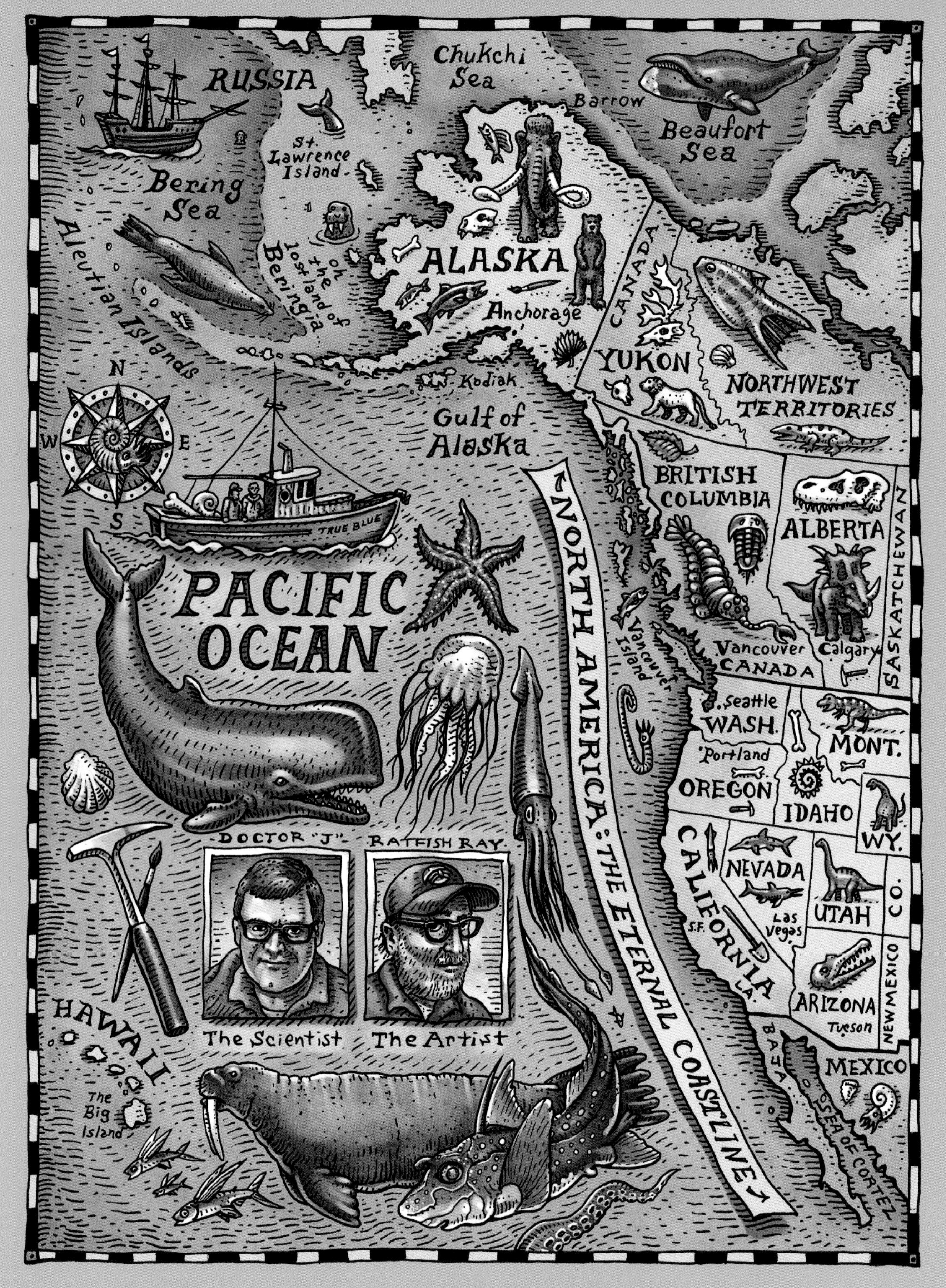
RUSSIA
Chukchi Sea
Barrow
Beaufort Sea
St. Lawrence Island
Bering Sea
Aleutian Islands
oh the lost land of Beringia
ALASKA
Anchorage
CANADA
YUKON
NORTHWEST TERRITORIES
Kodiak
Gulf of Alaska
N
W
E
S
TRUE BLUE
BRITISH COLUMBIA
ALBERTA
SASKATCHEWAN
PACIFIC OCEAN
NORTH AMERICA: THE ETERNAL COASTLINE
Vancouver Island
Vancouver
CANADA
Calgary
Seattle
WASH.
MONT.
Portland
OREGON
IDAHO
WY.
DOCTOR "J"
RATFISH RAY
CALIFORNIA
NEVADA
UTAH
CO.
S.F.
Las Vegas
L.A.
ARIZONA
Tucson
NEW MEXICO
HAWAII
The Scientist
The Artist
BAJA
MEXICO
SEA OF CORTEZ
The Big Island

slows down. A rising sea level will flood the coast, turning river valleys into estuaries and pushing beaches onto the shore. A falling sea level will expose old seafloor to the air and allow forests to colonize what used to be kelp beds.

When you combine all of these things, the result is that the relationship of the continent with its adjacent ocean is always changing. For as long as a coast exists, its mountains, hills, headlands, rivers, estuaries, bays, and beaches will slowly change their shape – even their location – because they are at the mercy of relentlessly moving geologic plates and an ever-changing sea level.

And while all of this geology is going on, climate changes, ecosystems form and disband, plants and animals live and die, and species evolve and become extinct. Fortunately, one of the most archival features of our planet is that it buries its dead (at least some of them). Animals

FOUR LAND MAMMALS LINKED TO THE SEA...
THE HORSE, THE BEAR, THE HIPPOPOTAMUS AND THE ELEPHANT

KANSAS HIPPY PAINTS FISH IN THE RAIN

and plants (or parts of them) occasionally get buried in sand or mud or lava or tar and become fossils. And because the coast is always moving up or down, and because the sea level is always moving up or down, and because the coastline is always eroding or depositing, fossils are continually being formed, deposited, uplifted, exposed, and eroded along coastlines. Coastlines are fabulous places to find fossils.

As one of the longest-lived coasts on Earth, the West Coast has seen a span of time that fully encompasses the evolution of complex life. Yet despite its great antiquity, it is still a geologically active place with coastal mountain ranges, raging rivers, earthquakes, faults, subduction zones, submarine canyons, volcanoes, landslides, and tsunamis. The northern section of the coast is home to endless fjords, huge ice sheets, sea ice, and mountain glaciers. This eternal coastline has been a stage for the evolution and extinction of myriad coastal animals, plants, and ecosystems. Put another way, this place has seen millions of variations on the theme of "surf and turf" over the last three-quarters of a billion years.

There are some groups of plants and animals whose stories are particularly well told by the fossils of the West Coast. Some of them, like redwood, ginkgo, and palm trees, live wholly on land; others, like marlin, ratfish, and squid and their extinct relatives, the ammonites, live only in the saltwater. Then there are the salmon who are born in fresh water and spend their lives at sea before returning to spawn in the streams of their birth, or the crabs that enjoy the saltwater at high tide only to scurry around on land at low tide. And finally, there are the animals that are amphibious in evolutionary time: the dugongs and sea cows who are related to elephants; the whales and dolphins whose relatives include hippos; seals, sea lions, and walrus who count the bears as their cousins; and finally, the extinct desmostylians (who appear throughout this book), whose ancestry allows them to be legitimately called sea horses.

Desmostylian

And then there are people. People have lived on the West Coast for the last 14,000 years or so, and the evidence of their existence is buried like the fossils before them. The past 500 years have seen the arrival of people from other continents, along with the expansion of cities, roads, and farms and the exploitation of forests, fisheries, and wildlife.

Our goal in this book is to travel the West Coast from Cabo San Lucas in Baja California to Prudhoe Bay, Alaska, and to tease out the stories of how it came to be the place it is today. I am a paleontologist and museum director who grew up in Seattle and experienced the West Coast as a kid at the same time that I was coming to love fossils. Ray is an artist who lives in Ketchikan, Alaska, who is known for his images of and obsession with fish. Back in 1992, Ray took a long road trip from Ketchikan to Kansas with the fish-inspired writer Brad Matsen – a trip that resulted in a book about the fossils, fish, and the ocean. Ray and I met shortly thereafter, and we did our own two-guy road trip between 2000 and 2005 to experience the fossils of the American West. That book, *Cruisin' the Fossil Freeway: An Epoch Tale of a Scientist and an Artist on the Ultimate 5,000-Mile Paleo Road Trip*, was so much fun that that we decided to take a run at the West Coast.

Over the last ten years, Ray and I have traveled together on a series of road trips, boat rides, flights, hikes, dives, swims, and museum visits to try and make sense of the long story of the West Coast. Both of us have both been inspired by the Native peoples, explorers, scientists, and artists who have inhabited, interpreted, and imaged aspects of this magnificent coastline. We have been impressed by the changes of the last few hundred years and the even more amazing changes of the last few hundred million years.

CALIFORNIA & NEVADA
OREGON
IDAHO
NEVADA
CA
AZ
ARIZONA
MEXICO
BAJA CALIFORNIA
Enhydra macrodonta oldest known sea otter
Thalatto-saur
Rustler Ranch Mastodon
Fagesia
Mount Shasta
Herpetocetus Dwarf Baleen Whale
Californosaurus
Hypsilophodontid
Shastasaurus
Eureka! Shell Fossils Galore
Fur Seal Callorhinus
Megamouth Shark
Hamiticeras
Large Eocene Leaves
Hesperornis
Macginitea
Artocarpus
Ancyloceras
Gateway Science Center
Desmostylus
Plesiosaur
Enaliarctos
Chalk Bluff Flora
Pteresaur
Sierra College Natural History Museum
C. megalodon
Tetrameryx irvingtonensis
Bison latifrons
Miocene Rhino
Frilled Shark
Mammoth Rubbing Rocks
Petrified Forest
Black Hawk Gomphotheres
Ice Age Horse
Megalonyx Ground Sloth
Univ. of CA Museum of Paleontology
Oakland Museum of California
San Francisco
Parabalaenoptera baulinensis
California Academy of Science
Flightless Auk
"The Boy Paleontologists"
Stanford Paleoparadoxia
San Jose
Children's Discovery Museum
"Lupe" the Mammoth
"Diaper Tuesdays"
Fossil Discovery Center of Madera County
Plotosaurus
Elephant Seal
Castroville
Monterey Bay Aquarium
Pacific Grove Museum of Natural History
ED RICKETTS
Mosasaur
Panoche Hills
Artichoke Mammoth
Big Sur
Sardine
Dusisiren jordani at Santa Cruz Museum of Natural History
Paleoparadoxia "Ancient Paradox"
Anchovy
Bonnerichthys
Giant Killer Dolphin Macrodelphinus
Parapontoporia wilsoni
Agatized Whale Brains!
Allodesmus
Food for 'Meg'
Giant Sea Cow Hydrodamalis cuestae
Pelagornis Giant Bone-Toothed Bird 17 ft. wings!
Cetotherium "Whale Beast"
Arctodus Short faced Bear
Santa Barbara Museum of Natural History
Point Conception
Platform Holly
Miocene Mola mola
Tiny Miocene Anglerfish
Gomphotaria
Love Lab
Prehistoric 'Killer' Sperm Whale
An Oarfish...
Acentrophryne sp.
from L.A. County
As compared to a modern dog...
Channel Islands Pygmy Mammoth
Pelagiarctos
Pachydiscus a big one, near L.A.
Dusignathus seftoni four tusked walrus
Valenictus chulavistensis Walrus from Chula Vista
Gomphotaria pugnax Giant Four Tusked Walrus
Ammonite
Smilodon
Columbian Mammoth
American Lion
Peraceras Rhino
Moropus
Helicoprion
Eocene Leaves
American Cheetah
Permian Ammonoids
Our Lobefin Kin!
Cymbospondylus
Belemnite
Gymnotoceras
Tinirau Half fish, Half man
Asterolepis
Meekoceras
Giant Rhizodont Lobefin
Lamiaspis
Devonian Pteraspid fish
The oh so mysterious Omphalosaurus
Loneliest Road
Reno
Large Lichid Trilobites
Shonisaurus
Nevada State Museum
Berlin Ichthyosaur State Park
Pirumaspis
Canadaspis
Cheirolepis
Machairodus
Gomphothere
Miocene Leaves
Lunar Crater National Natural Landmark
Ottoia
Teleoceras
Tuzoia
Anomalocaris
Mastodon - found by teens near Gardnerville
Extraterrestrial Highway
Olenellus
Esmeraldina trilobite
Zacanthoides trilobite
Alien enough for ya?
Death Valley National Monument
AREA 51
A big Camel
Giant Sloth
Kings Canyon National Park
Dunkleosteus
Saurolophus
Sequoia Nat. Park
Tule Springs Ice Age Fossil Beds
Las Vegas Natural History Museum
Zabriskie Point
LAS VEGAS
Hoover Dam
Bison antiquus
Anomolocaris
Megalodon
Sharktooth Hill
Buena Vista Museum of Natural History
Dinosaur Tracks
Gertie the Giant Ground Sloth
Capricamelus gettyi Short-legged Goat-Camel
Bakersfield
America's Oldest Elephant
Zygolophodon
Phenacodus Paleocene Mammal
Amphicyon Bear Dog trackway
Bristolia insolens Trilobite
Smilodon
Alf's Pivotal Peccary
Harlan's Ground Sloth
Barstow
Oreodonts
Peachella iddingsi Trilobite
Mojave Desert
Raymond M. Alf Museum of Paleontology Claremont, CA
"The Flames of Enthusiasm"
George C. Page Museum
La Brea Tar Pits
HOLLYWOOD
San Bernardino County Museum
"Dinny"
Los Angeles
Natural History Museum of Los Angeles County
Cabazon "Claude Bell's Dinosaurs"
Joshua Tree National Park
Incredible Wind God Bird Aiolornis incredibilis
Dire Wolf
Cooper Center
Aquarium of The Pacific
Western Science Center
Giant Camel Gigantocamelus
Imperial Desert Walrus
Saber Toothed Salmon
Parvicornus Brontothere
Salton Sea
Anza-Borrego Desert State Park
Valenictus imperialensis
Pachydiscus
California's most complete dinosaur
Ricardo Breceda Sculptures
Sespia Oreodont
San Diego
San Diego Museum of Natural History
Tessorolax
Aletopelta
Tijuana
Calexico
Yuma
Jalisco
Magnapaulia
Colorado River

# CALIFORNIA

# 1 TOUCHED BY THE TAR

Los Angeles has always been a mixed bag for me. My parents met and dated here during the city's salad days of the 1950s. As a kid, I can remember visiting relatives and seeing endless orange orchards, 23-cent-a-gallon gasoline, and a very fresh Disneyland. As I grew up and as LA grew out, the vision of paradise smogged over and LA became a little less attractive in my mind. I can remember flying into Burbank in 1992 during the height of the riots and watching SWAT teams collecting their assault rifles at the baggage claim. Twenty years later, I met Ray in the baggage claim at LAX on a Sunday morning. He had flown in from Seattle; I had come from Denver. We had decided to start our road trip in the City of Angels.

The same things that had attracted my parents to Los Angeles also attracted a lot of other people, and the LA Basin grew and grew and grew, eventually becoming the endless America mega-city that it is today. There are now more than 12 million cars and 20 million people in the LA Basin, and it has an astounding acreage of highway surface. Los Angeles represents one view of the

A pair of saber-toothed cats and a pack of dire wolves stalking Columbian mammoths and giant ground sloths, mired in the tar seeps of ancient Los Angeles.

future of the world, and given the movie industry's love of apocalypse, we have all seen many versions of the Los Angeles of the future. Ray and I decided to rent a hybrid.

It was a sunny, beautiful Sunday morning, and there was no traffic. We easily navigated our way to Culver City and found a vegan café where we ate fake fish tacos and relaxed in the sun.

We were headed for the La Brea Tar Pits, the world's largest and most famous Ice Age fossil site, which just happens to be located in downtown Los Angeles next door to the Los Angeles County Museum of Art.

Tar does not seem to be a natural thing, but it is. The sticky black stuff used to hold gravel together to make blacktop roads is nothing more than petroleum (commonly called oil) that has lost some of its liquidity through evaporation. Petroleum, of course, is the source of gasoline and thus the founding liquid for our car culture. The ultimate source of petroleum is a mixture of microscopic marine plants and animals – plankton by another name – that die and settle to the bottom of the sea where they are covered by mud and eventually find themselves deeply buried. When the burial is deep enough, the pressure and the temperature grow to the point that the buried organic-rich mud enters what is known to geologists as the "oil window." In this zone, the conditions are right for the cooking of buried plankton into petroleum. Once the petroleum forms, some of it stays where it is, and some of it begins to move – sometimes along cracks, and other times in the space between sand grains. This movement is facilitated by heat, gravity, and pressure changes beneath the surface.

Oil can make its way to the surface by natural processes as well. In some places, it simply seeps to the surface where the volatiles evaporate, leaving the sticky stuff known as tar. If it evaporates even more, it becomes a solid black mass called asphalt. This process has been happening in downtown Los Angeles for a lot longer than there has been a downtown Los Angeles – at least for the last 40,000 years.

And it turns out that you don't want to step in tar. In fact, if you step in tar that is more than a few inches deep, you will be unlikely to get your foot out without help. For this reason, tar seeps are surprisingly dangerous places. They form natural traps where innocent animals can get stuck and never get unstuck.

The tar seeps at La Brea were first noted for their economic potential, and by the 1870s, the Hancock family, who owned the more than 4,400-acre Rancho La Brea, was mining asphalt and selling it as a binder to hold cobblestones together. The mining of the asphalt created pits in the ground, and some of these pits filled with water and became small ponds. As they mined the asphalt, the Hancocks noticed that it contained an awful lot of bones.

It didn't take people long to realize that the bones belonged to animals that had been trapped by tar, had died, and had been buried in a mixture of sand and silt. It was the perfect fossil-making machine: An animal would simply get stuck and die. Then a flood from a nearby stream would bury it in silty mud. Then the tar would seep up through the sediment and trap another animal, and after a while there would be a sticky black stack of tar and mud and skeletons.

### That Bubblin' Crude

**The very first oil well was drilled in Pennsylvania in 1859. Back then people called it rock oil, both because its origins were not entirely clear and to distinguish it from whale oil. When an oil well was drilled into a zone of intense pressure, the result was a gusher – an oil well that literally sprayed oil high into the sky until it could be capped and controlled. Some of the earliest oil exploration and production on the West Coast happened in the Los Angeles Basin, and even today, several billion barrels of oil lie beneath the city's surface. You can even see pumps bringing up oil from sites near Beverly Hills High School. The *Beverly Hillbillies* was an interesting 1960s mash up of some country folk somewhere who struck it rich on some "bubblin' crude" and moved to the Beverly Hills. Turns out, there was oil there too.**

And it was even better than that. Once an animal was trapped, it was available for predators or scavengers who are always on the lookout for easy pickings. They would move in for an easy meal and get trapped, their bodies luring more predators. Paleontologists call this a "predator trap" and are delighted with the resulting bounty of carnivores.

For Ray and me, the concept that the same petroleum that gave us the Los Angeles freeway system also gave us the world's finest Ice Age fossil site is pretty ironic. We walked into the Page Museum full of anticipation and were greeted by Shelley Cox, the chief preparator and queen of the tar. She is a small, dark-haired woman who is fully devoted to the cause. She had started vol-

A MERE 14,000 YEARS AGO, LOS ANGELES WAS FULL OF GIANT SLOTHS, WOLVES, SABER-TOOTHED CATS, AND MAMMOTHS.

unteering at La Brea in 1973 and was hired two years later. The Page Museum opened to the public in 1975, and she worked there right up until her recent retirement. Shelley presided over a team of forty-two very focused volunteers who use needles, brushes, solvents, blades, and trowels to extract fossils. When we met, she quickly set about correcting our misconceptions: "It's not tar, it's oil and there never were pits." So much for the La Brea Tar Pits. I glanced over at Ray, and I could see him processing the concept of the La Brea Oil Seeps.

In 2008, the nearby art museum had excavated an underground parking garage, and during their digging they had come across bones. In California, federal, state, and, in some cases, county laws stipulate that contractors hire trained paleontologists to observe active excavations and salvage the fossils for science. The art museum solved its bone problem by digging around the tar-soaked zones and then wrapping them with boards and cables. Then they hoisted these giant tar-infused sand blocks out of the construction zone and over to the backyard of the Page Museum. Shelley called it Project 23, as it had twenty-three of these giant fossil-filled monoliths to process. Block 1 weighed 56 tons!

We went out to look at the blocks and ran into two of Shelley's volunteers, Karin Rice and Carrie Howard. Karin was the daughter of one my mom's Los Angeles roommates. I had known that she was a salvage paleontologist but not that she was into the tar, and I was surprised and delighted to see her. Carrie had started volunteering a few years earlier and had "fallen in love with tar fossils." Both volunteers were covered with asphalt and seemed really happy. Carrie said, "I

Shelley Cox gazing lovingly at the skull of *Panthera atrox*, the American lion.

love dirt." They were digging on the top of Block 1 and were uncovering a jumble of bones, including identifiable parts of a *Panthera atrox* (American lion), a *Smilodon fatalis* (saber-toothed cat), a bison, and an eagle-sized bird. They were happy to show us what they were doing and just as happy to get back to it.

Trevor Valle and his *Smilodon* tattoo.

Walking back to the museum, we stopped by Pit 10, which was a 12-foot-wide pool of bubbly oil. I stopped to take a picture of Ray posing like he was stuck in the tar, and when I tried to walk I discovered that my own feet actually were firmly stuck. Shelley said that 6 inches of oil would trap an animal of any size. Ray and Shelley pulled me free before the scavengers set in, and we went into the museum and entered the lab space, which is open to view for museum visitors.

The art museum project had been a big boon for Shelley's team, and they had recovered a nearly complete mammoth they named Zed, whom volunteers were busily exposing. They also had a partial camel named Clyde, a lion named Fluffy, and a saber-toothed cat named Bagheera. Ray was immediately drawn to a giant red-headed preparator named Trevor Valle who sported a big red beard and a nearly life-size tattoo of a *Smilodon* skull on the front of his lower leg. We were rapidly getting the picture that the volunteers at the Page were a very committed bunch.

Over the years, La Brea has yielded literally millions of bones. The University of California at Berkeley worked here in 1911–1912; the Los Angeles County Museum of History, Science, and Art started a two-year dig in 1913; and Chester Stock at the California Institute of Technology took over after that and continued digging until 1950. The bones range in age from recent victims like squirrels and birds to critters that lived there as long as 40,000 years ago. The list of bod-

TOUCHED BY THE TAR

ies is an amazingly complete accounting of plant and animal species and includes one human – a 9,000-year-old skeleton of a twenty-year-old woman.

The site truly is a predator trap, and more than 70 percent of the bones come from predators or carnivores. Since predators are much less common than herbivores, this ratio shows how effectively the trap worked. Paleontologists can match up bones and determine the "minimum number of individuals" represented. Using this method, they have documented that La Brea has yielded fossils of at least 3,000 dire wolves and 1,600 *Smilodon.*

Shelley took us back into the collections and showed us drawer after drawer of isolated sabers and case after case of bones from dozens of different species. Other carnivores included lots of coyotes, about sixty American lions, and about thirty short-faced bears (*Arctodus*). The American lions and short-faced bears were both very large-bodied carnivores. The lion was larger than an African lion, and the short-faced bear was considerably larger than a grizzly bear. *Smilodon*, with its big sabers, gets most of the press, but clearly Los Angeles was a very dangerous place 12,000 year ago.

We wandered into the exhibits and saw amazing mounts of camels, horses, bison, mammoths, mastodons, short-faced bears, American lions,

Isolated *Smilodon* saber teeth are surprisingly common at La Brea.

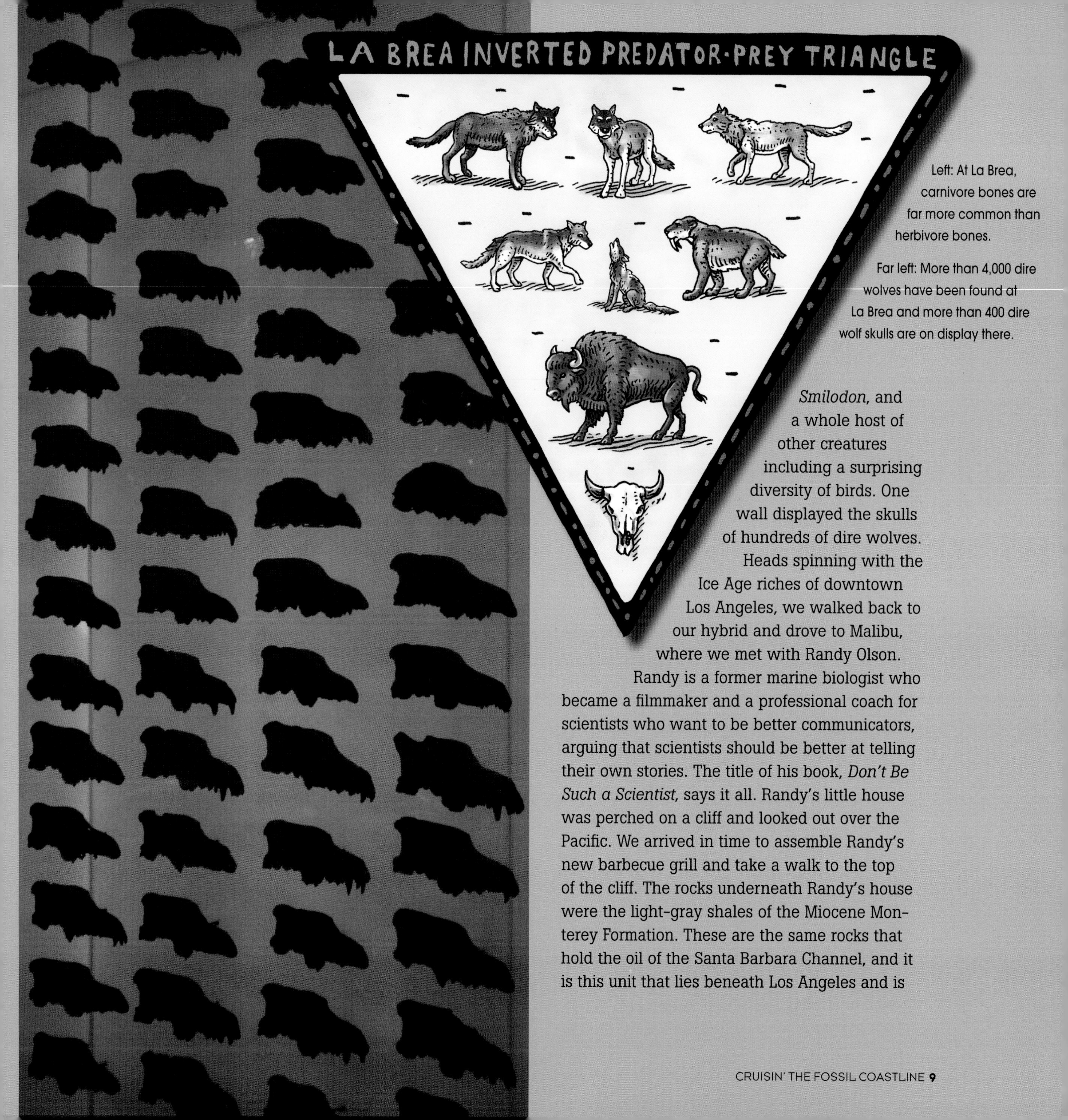


Left: At La Brea, carnivore bones are far more common than herbivore bones.

Far left: More than 4,000 dire wolves have been found at La Brea and more than 400 dire wolf skulls are on display there.

*Smilodon*, and a whole host of other creatures including a surprising diversity of birds. One wall displayed the skulls of hundreds of dire wolves. Heads spinning with the Ice Age riches of downtown Los Angeles, we walked back to our hybrid and drove to Malibu, where we met with Randy Olson.

Randy is a former marine biologist who became a filmmaker and a professional coach for scientists who want to be better communicators, arguing that scientists should be better at telling their own stories. The title of his book, *Don't Be Such a Scientist*, says it all. Randy's little house was perched on a cliff and looked out over the Pacific. We arrived in time to assemble Randy's new barbecue grill and take a walk to the top of the cliff. The rocks underneath Randy's house were the light-gray shales of the Miocene Monterey Formation. These are the same rocks that hold the oil of the Santa Barbara Channel, and it is this unit that lies beneath Los Angeles and is

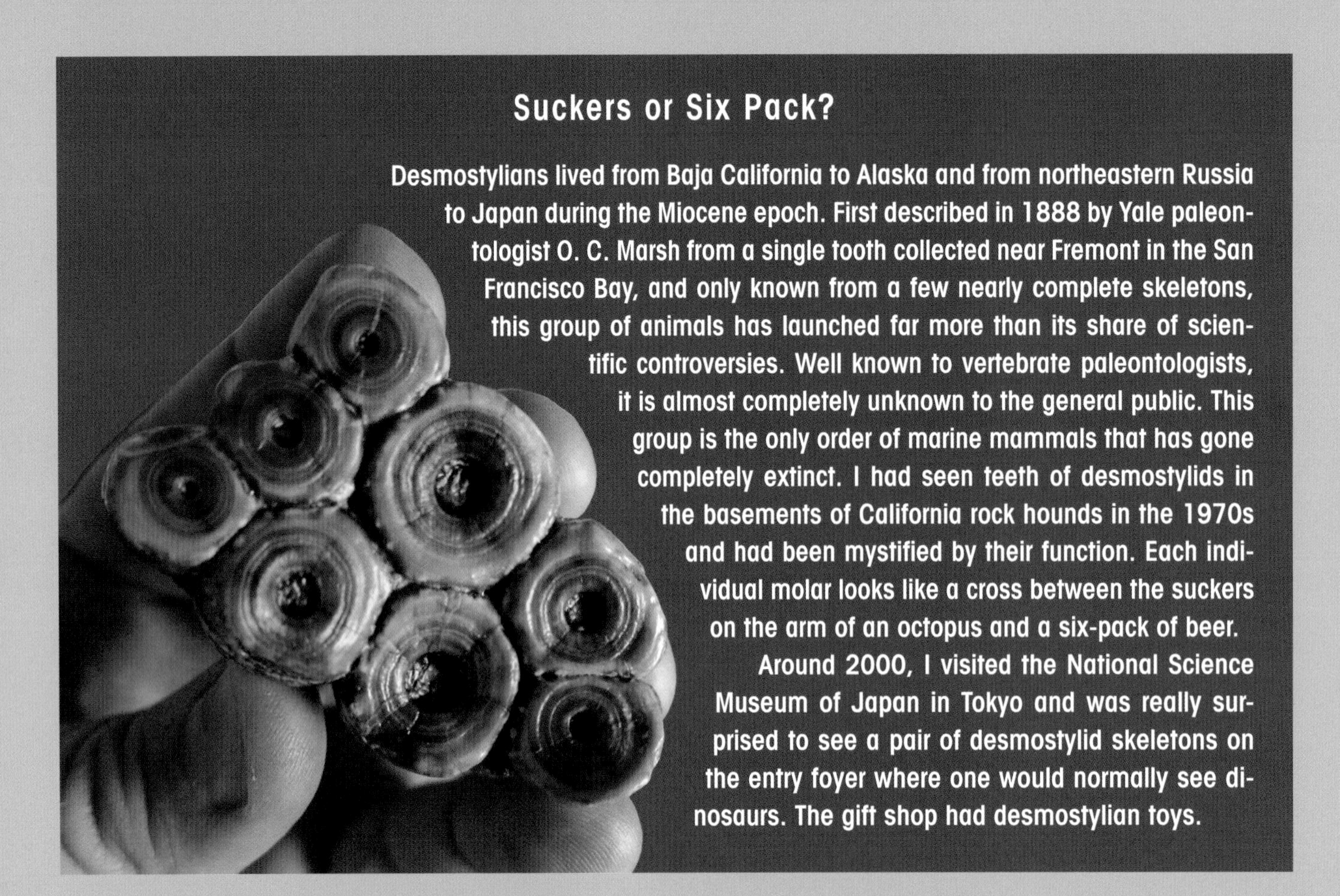

### Suckers or Six Pack?

Desmostylians lived from Baja California to Alaska and from northeastern Russia to Japan during the Miocene epoch. First described in 1888 by Yale paleontologist O. C. Marsh from a single tooth collected near Fremont in the San Francisco Bay, and only known from a few nearly complete skeletons, this group of animals has launched far more than its share of scientific controversies. Well known to vertebrate paleontologists, it is almost completely unknown to the general public. This group is the only order of marine mammals that has gone completely extinct. I had seen teeth of desmostylids in the basements of California rock hounds in the 1970s and had been mystified by their function. Each individual molar looks like a cross between the suckers on the arm of an octopus and a six-pack of beer.

Around 2000, I visited the National Science Museum of Japan in Tokyo and was really surprised to see a pair of desmostylid skeletons on the entry foyer where one would normally see dinosaurs. The gift shop had desmostylian toys.

the source of the La Brea oil. The Monterey itself is a geological formation that was deposited in shallow seas and as a result also has lots of marine fossils.

The next morning, we headed into the Natural History Museum of Los Angeles County (NHM-LAC is such an unwieldy acronym that we'll use NHM) in Exposition Park. This venerable institution opened to the public in 1913 and has been doing and sharing science ever since. This is a big museum and we had a very full dance card. For a paleontologist, going to a big museum is as complicated as going to a family reunion. Obviously there are the exhibits, but there are also the scientists, administrators, educators, and collections. All of them have fascinating tales to tell, and we wanted to hear them all.

For Ray, the NHM is a very special place. It was here back in the late 1980s that he got his first glimpse of *Helicoprion*, the buzz saw shark. This glimpse turned into a full-blown obsession that has taken years to subside. Fortunately, there were new distractions. The museum was about to open its new mammal hall, and we were able to get a sneak peek tour of the construction site. The exhibit (that has since opened, along with a magnificent dinosaur hall) contained a mixture of fossil and living marine mammal skeletons and mounts. Two suites of skeletons stood out: the Ice Age mammals from La Brea, and the Oligocene to Pliocene marine mammals from several spots in California. Standing next to an impressive mounted skeleton of an American lion from La Brea, Ray looked like a chew toy. Our guide told us of an ongoing squabble about this extinct carnivore, and how some scientists argue that it is not a lion at all but rather a huge jaguar. The concept of a massive mega-jag was truly terrifying.

For nearly four decades, Larry Barnes had been the curator of fossil marine mammals at the NHM. Barnes is renowned as a brilliant scientist but one who keeps his cards close and his specimens and localities even closer. I had been hearing stories of Larry Barnes my whole fossil-finding life, but I had never met him. Now we had the first view of his career-capping exhibit.

The fossils did not disappoint. For me, the center of attraction was a huge 11-million-year-old skeleton that had been collected in a construction site at a place called Aliso Viejo in Orange County. Maybe 12 feet long, it was nearly complete and bore the self-descriptive name *Neoparadoxia*, the "new paradox." This animal is from an extinct lineage of marine mammals known as desmostylians.

Larry's *Neoparadoxia* was a dream come true. Here was a hippo-sized marine mammal with broad feet and wide toes. Larry's mount shows it in an open swimming posture, and I really wanted to see what this animal looked like in life. Fortunately, I had an artist. I set about with my plan to dislodge Ray's *Helicoprion* obsession and replace it with my desmostylian obsession. My plan worked beautifully, and Ray was sucked into the world of extinct marine mammals.

In addition to the desmostylian, there were several other amazing marine mammals: a pair of Miocene *Dusisiren* sea cows from the hills above Santa Cruz, a complete Miocene pseudo-sea lion (*Allodesmus*), and the skeleton of a

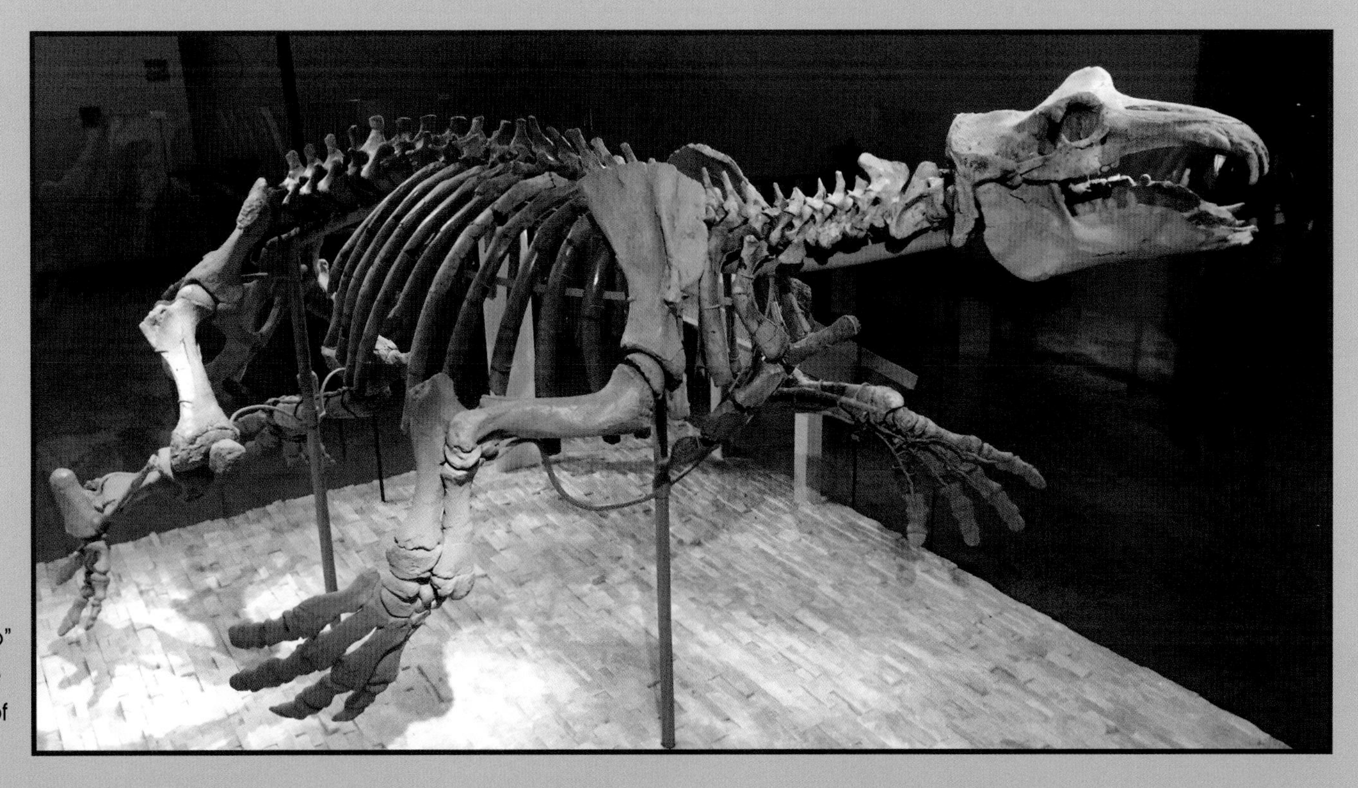

The amazing *Neoparadoxia cecelialina* "Desmo" skeleton on display at the Natural History Museum of Los Angeles County.

small sperm whale (*Aulophyseter morricei*) – the latter two from the dusty hills outside of Bakersfield. It was clear that Larry was beginning to play his cards.

Ray and I requested an audience with the man. It was granted. We were shown to his office and took seats. Larry was long-limbed and had a full head of gray hair – he reminded me of General George Custer. His body language suggested that he didn't quite know what to make of a paleobotanist-artist pair who were deeply interested in marine mammals. He was flanked on both sides by collection managers Sam McLeod, mammal man, to his right and Gary Takeuchi, fish man, to his left. Mammal paleontologist Xiaoming Wang watched from the wings. For the next three hours, we asked questions, and Larry dipped into his deep well of West Coast marine mammal knowledge.

Because much of the West Coast is vegetated (as opposed to the deserts and badlands of the American West), paleontologists are generally limited in their access to fossil-bearing bedrock. Coastal cliffs, wave-cut terraces, and rocky beaches provide some of the best outcrops. River cuts are another place where fresh bedrock is actively being eroded. Since coastal cliffs and beaches are always actively eroding and collapsing, new fossils are constantly being exposed and you have to be at the right place at the right time to discover them. The successful fossil hunter usually finds a promising spot and visits it repeatedly, often after major storms or rainfalls that may have caused landslides and slope failures. Although there are not too many knowledgeable fossil finders, they nevertheless tend to be secretive about their spots and competitive about their timing, often hitting the beaches just after dawn to have the first look.

In California, where economic growth has been booming for more than a century, construction sites and road building exposes lots of bedrock, and with it, lots of fossils. The combination of the rapid growth and the well-enforced California laws requiring trained paleontologists on-site has yielded a bonanza of new fossil finds.

Larry's *Neoparadoxia* was found during construction in Orange County when an earthmover sliced into its right side; fortunately, the salvage paleontologist was able to prevent further damage and retrieve the skeleton. It was found in the Monterey Formation, the same layer of rock that lay beneath Randy Olson's home in Malibu and was the source for the La Brea oil. In the Orange County site, the formation was about 10 to 12 million years old and also preserved fossils of whales, dolphins, birds, fish, and turtles.

One of the tricky things about studying the animals of ancient coastlines is that the bodies of land animals and plants often wash out to sea, so their fossils are found comingled with marine organisms even though they did not live with

Larry Barnes and the skull of *Enaliarctos*, one of the earliest known pinnipeds.

them. This is pretty easy to sort out for animals that are clearly land dwellers, such as camels, but much more difficult for extinct beasts with unusual anatomy. Larry was pretty convinced that the *Neoparadoxia* had a hippo-like lifestyle, swimming in coastal rivers, drinking fresh water, and feeding on land plants.

Like many vertebrate paleontologists of the West Coast, Larry had done his graduate studies at the University of California at Berkeley. He focused on West Coast marine mammals, and his interests ranged from California to Japan. He

## An Abundance of Riches

One of the most simultaneously fascinating and frustrating things about museums is that they have so many more extraordinary items behind the scenes in the vast research collections than they can ever display. Across the wide field of museums, there is fortunately a growing trend of digitizing collections to counter this problem and to make collections accessible to everyone.

had a wide array of contacts, and when someone would find a fossil marine mammal somewhere on the West Coast, the specimen would often make its way to Larry's lab in Los Angeles. This is a pretty common practice in a world where fossils are widespread and experts are not. Many of the marine mammal fossils from the Burke Museum of Natural History and Culture in Seattle, for example, had been loaned to Larry, and some of them had been in his in-box for decades.

It rapidly became apparent to Ray and me that we were in the presence of a huge amount of knowledge. Larry had roamed far and wide, and for many years he had been the go-to man when a fossil marine mammal turned up.

Our conversation eventually turned back to the Monterey Formation and the fossils found there. Gary Takeuchi described the discovery of deep-water anglerfish fossils at the construction site for a new Los Angeles subway stop at the corner of Vermont and Wiltshire. Today, these fish live at depths of 3,000 feet, so their discovery above sea level in Los Angeles says a lot about how the southern California landscape has moved over time. It was becoming pretty clear to us that the fossil record of California had vast amounts of unrevealed detail.

Larry pulled some drawers and showed us the tip of a fossil iceberg. One specimen that stuck in my mind – and stuck hard – was a fossilized whale brain from a site near San Luis Obispo. I would never have thought it possible that a brain could fossilize, but here, in a fossil whale skull, was an unambiguous agatized brain. I have never seen anything like it, before or since.

Larry ended our session by telling us a few stories about famous California paleontologist Chester Stock. Stock devoted his life to the vertebrate paleontology of California, working extensively on the La Brea mammals but also digging dinosaurs and marine reptiles. He had a reputation as a superb lecturer and a brilliant scientist, and we would see his name occur again and again in our travels. Larry intimated that Stock was even more interesting; rumor had it that he could make books fly off the shelf.

Next stop was my old friend, Argentine dinosaur paleontologist Luis Chiappe. Like me, Luis was beginning to move up the ranks of museum management, so we had lots of notes to share. Luis was busily working on the renovation of the NHM dinosaur halls and had taken the opportunity to launch a series of expeditions to Colorado and Montana in search of new skeletons to add to the

Luis Chiappe holding a cast of the skull of *Carnotaurus*, a Patagonian theropod dinosaur.

ISOPODS, IPOD AND I ...

museum's already rich holdings. Ray and I were curious about California dinosaurs, and it turns out that there are enough Mesozoic reptiles in California to fill a book. That book, *Dinosaurs and Other Mesozoic Reptiles of California* by Dick Hilton, recounts a long history of small and scattered finds from nine different counties, adding up to a pretty amazing selection of Cretaceous creatures. A lot of these finds were made by amateur collectors, and Dick does a nice job of giving them credit for their finds.

In fact, California's first dinosaur was found in 1936 by a seventeen-year-old kid named Allan Bennison, who had ridden his bike 36 miles to hunt for fossil shells in the Panoche Hills west of Fresno. This was close to home for me; my dad was born in Fresno in 1932 and may have heard of the discovery in his local school. Allan's discovery of a partial duck-billed dinosaur skeleton opened a new chapter in California paleontology, and he soon caught the attention of paleontologists in both Berkeley and Los Angeles. He followed this find with discoveries of marine reptiles and went on to have a long career as a geologist.

Allan's discoveries increased professional interest in the Panoche Hills, and a number of subsequent expeditions resulted in some remarkable finds. One dinosaur that caught my interest was *Saurolophus*, a crested hadrosaur known also from Montana and Alberta, Canada. In fact, Chester Stock and his crews had collected two hadrosaur skeletons in 1939 and 1940. To me, one of the very interesting things about *Sauro-*

*lophus* is that I had seen its fossils in Mongolia when I was there in 1997. This means that there must have been some connection between North

America and Eurasia as long as 70 million years ago, and that our trip up the West Coast would be following a very old trail indeed. Luis and his team eventually renamed this dinosaur *Augustynolophus morrisi*, and, in 2017, it was declared the official state dinosaur of California.

The next day, we found our way to an offsite storage facility in East Los Angeles where the NHM keeps its vast holdings of invertebrate fossils. In a nondescript warehouse, we were confronted with a collection of more than 17 million fossils, most of them from the West Coast. Here was a trove of fossils that anchored much of what we know about the marine paleontology of Oregon, California, and the Baja Peninsula. For a guy that loves ammonites and fossil crabs, I had found my *Raiders of the Lost Ark* moment. We spent hours pulling drawers, inspecting fossils, and time traveling to ancient shorelines. During the riots of 1992, this facility had been right in the middle of the troubles, and the collection managers had boarded up the windows and stayed in the building to protect the fossil treasures.

It was clear that we could spend weeks behind the scenes at the NHM, and it was also clear that the staff were doing a great job of building new exhibitions. When their fossil mammal and dino-

Above: Harry Filkorn and Lindsey Groves with northern California ammonites.

Right: Kirk fell in love with this *Eupachydiscus lamberti* ammonite from Los Angeles County.

saur halls opened a few years later, they received great acclaim from the museum community for putting actual California fossils on display. For the first time, Californians could access desmostylians, California dinosaurs, and a whole lot more.

Next, we headed up the coast toward Santa Barbara. Along the way we passed the town of Carpinteria, which has its own oil seeps, that had preserved fossils like a miniature La Brea. Back in the 1930s, the site had yielded a mastodon.

One of our first destinations in Santa Barbara was offshore. I had recently met the owner of one of the main oil-drilling platforms in the Santa Barbara channel, and he had invited me to inspect the rig. Ray and I showed up at the dock and decked out in safety suits and orange vests. Ray was informed that no beards were allowed on the platform – they interfered with the respirator masks and everyone on the platform had to be ready to don a mask at a moment's notice. We hopped onto the crew transport vessel and took the twenty-minute ride out to the platform. Along the way, we passed several natural submarine oil seeps, places where petroleum was bubbling up from the seafloor. We could recognize them because of the rainbow-hued oil slicks that formed smooth patches on the sea surface. There is a tendency to think of petroleum as an artificial or industrial chemical, but it is truly a product of natural geological processes. It was becoming quite clear that this part of southern California had a very long and very natural relationship between petroleum and people. Off in the distance, we could see three of the largely uninhabited Channel Islands that lie to the south of Santa Barbara. Archaeological artifacts from these islands have shown that Native peoples used naturally occurring asphalt for a variety of purposes.

Approaching Platform Holly

Platform Holly was installed in 1966 and has been producing oil and gas since then. From the single platform that sits in 211 feet of water, numerous wells poke some 8,000 feet down into the Monterey Formation, those same rocks that are underneath La Brea and Randy Olson's house. The platform produces about 2,500 barrels of crude oil every day, which is then pumped through pipelines to processing plants on the Santa Barbara shore.

## Steller's Sea Ape

While on the crew ship, Ray had a very different conversation with Captain Pettit. The two shared an earnest discussion about the observational skills of Georg Steller, the eighteenth-century naturalist for whom Steller's jays, Steller sea lions, and the extinct Steller's sea cow are all named. Apparently, Steller once had a sighting of a very curious animal that he described as a sea ape. He knew sea otters, and this was no otter. It was a hairless, human-sized animal with flippered hands and feet and a neck that could swivel like a human's. Steller was an excellent naturalist, so his observations were taken seriously – but no one else has ever seen a sea ape. No one, that is, except Captain Pettit. When I sidled up next to Ray, he was drawing a picture of the sea ape under Pettit's supervision. To me this story was about as believable as an aquatic Sasquatch, but Pettit was firm in his belief that he had seen one as well. I told Ray that next time he should shave his beard.

DOG PADDLING LIKE CRAZY...
IT TURNED AND "SMILED" AT
US, AS IF TO SAY "YOU'LL
NEVER CATCH ME", AND THEN IT DISAPPEARED.

As we pulled up to the rig, Ray decided that his beard was more important than the visit, and he decided to stay on the crew ship with its captain, Darryl Pettit. I boarded the platform and got a great tour of the rig and a very deep sense of how profound safety consciousness pervades life out there on the platform. As it happened, we were visiting just a few days after the BP *Deepwater Horizon* rig had exploded in the Gulf of Mexico. Operating less than a mile from shore and in only 200 feet of water, Platform Holly was like a miniature version of the giant Gulf rig that was operating in more than 8,000 feet of water. The visit gave me a vastly increased appreciation for the challenges that have to be overcome to fill my gas tank. In 2017, the platform's operating company began the process to decommission the platform.

The Santa Barbara Museum of Natural History is located on a lovely sprawling campus high above the city. My friend Karl Hutterer was the director of the museum, and I had known him from Seattle where he used to run the Burke Museum. We went to the museum to see the fossils from both the Monterey Formation and the Channel Islands. In the places where the Monterey Formation is exposed on land, it crops out as a strange bright-white and lightweight rock known as a diatomite. Diatoms are the marine plants whose degradation

Skeleton of the pygmy mammoth (*Mammuthus exilis*) collected on Santa Rosa Island in 1994 on display at the Santa Barbara Museum of Natural History.

**In 2015, I finally had the chance to visit the fossil sites at Santa Rosa, and what I saw there was amazing. The bluffs along certain stretches of the coastline are studded with middens containing the remains of millions of crabs, clams, abalones, snails, whales, seals, and fish. The dates on the bones found by Orr are still under debate, but it looks as if some of those people may have been living there more than 11,000 years ago. The pygmy mammoths are slightly older but are very close in age, so it is unclear if people ever shared the island with the midget mammoths. Nevertheless, these sites are very strong evidence that some of the earliest Americans traveled along the coast from Alaska down to California, enjoying many a mixed seafood platter.**

forms the stuff of oil. A diatomite forms when these microscopic marine plants are dying and raining down on the seafloor at a place where there is no silt, sand, or clay. The resulting rock is composed entirely of diatoms. And, as it happens, these rocks can be very nice for the preservation of larger fossils. A diatomite pit near Lompoc, California, is famous for its fossilized schools of herring, while other have yielded crabs, hatchetfish, giant stingrays, sunfish, huge leatherback turtles, cormorants, early puffins, sea lion–like walrus (*Imagotaria*), early fur seals (*Pithanotaria*), sperm whales, and baleen whales (*Megaptera miocaena*). In 1955, the Antolini and Sons Quarry in Santa Maria produced an extraordinary fossil: the skeleton of a bird with a 17-foot wingspan that was named *Pelagornis orri* after the Santa Barbara Museum archaeologist Phil Orr.

Orr was known for his pioneering work on the Channel Islands. He had a remote field camp on Santa Rosa Island where he found the remains of some of the oldest humans ever discovered in North America. On the same island, he found the remains of pygmy mammoths. Apparently, even at the lowest sea level of the last Ice Age, the island was still more than 7 miles from the mainland. This meant that these mammoths had to swim at least that far to colonize the island. In 1994, a park ranger on Santa Rosa found the nearly complete skeleton of one of the pygmy mammoths,

COLUMBIAN MAMMOTH
ASIAN ELEPHANT
AFRICAN BUSH ELEPHANT
FOREST ELEPHANT
WOOLLY MAMMOTH
ELEPHAS
LOXODONTA
PRIMELEPHAS
GOMPHOTHERIUM
PLATYBELODON
ELEPHANTIDAE
STEGODON
AMEBELODON
GOMPHOTHERIIDAE
DEINOTHERIUM
MAMMUTIDS
MASTODON
DEINOTHERES
BARYTHERIUM
MOETHERIUM
STELLER'S SEA COW
MANATEES
DESMOSTYLIANS
PROBOSCIDEA
SIRENIA
DUGONGS
HYRACOIDEA
HYRAX
TETHYTHERIA
MAMMALIA

THE ELEPHANT FAMILY TREE

and Larry Agenbroad from the Mammoth Site in South Dakota came to excavate the fossil, which turned out to be 98 percent complete.

As Ray and I perused the collection of mammoth bones at the museum, we realized that not only were there pygmy mammoths on the island, but there were full-sized mammoths as well. Apparently, the pygmies evolved on the island from large stock that continued to migrate to the island from the mainland. Imagine showing up in a place where everyone looks like you, but only half as big.

During this trip, people kept telling us about a huge Indiana Jones–like warehouse somewhere in Orange County that housed endless fossils. On a solo trip (Ray couldn't make this one) to Los Angeles two years later, I made some calls, rented a car, and drove to Santa Ana. By this time the Cooper Center, a cooperative effort between the Orange County Park System and Cal State–Fullerton had formally opened. I was met by Meredith Rivin, the supervisor of the site. She was delighted to have visitors, and I was surprised to see what was stashed behind the nondescript doors of the industrial warehouse. Orange County is home to both marine and terrestrial rocks – and lots of construction. The active construction sites in the area have, since the 1970s, produced countless fossils that have been carefully monitored by county-mandated professional paleontologists. The many fossils they collected in Orange County ended up in this odd facility that was all collection and no museum. Once again, I was struck by how many fossils are found in California and how little anybody there knows about them.

Meredith Rivin with the skull and jaws of Waldo the walrus at the Cooper Center in Santa Ana.

Meredith was happy to show me her prizes, and as a big fan of walrus, I was delighted to see the remarkably complete skeleton of Waldo, a large Mio-Pliocene imagotarine walrus from

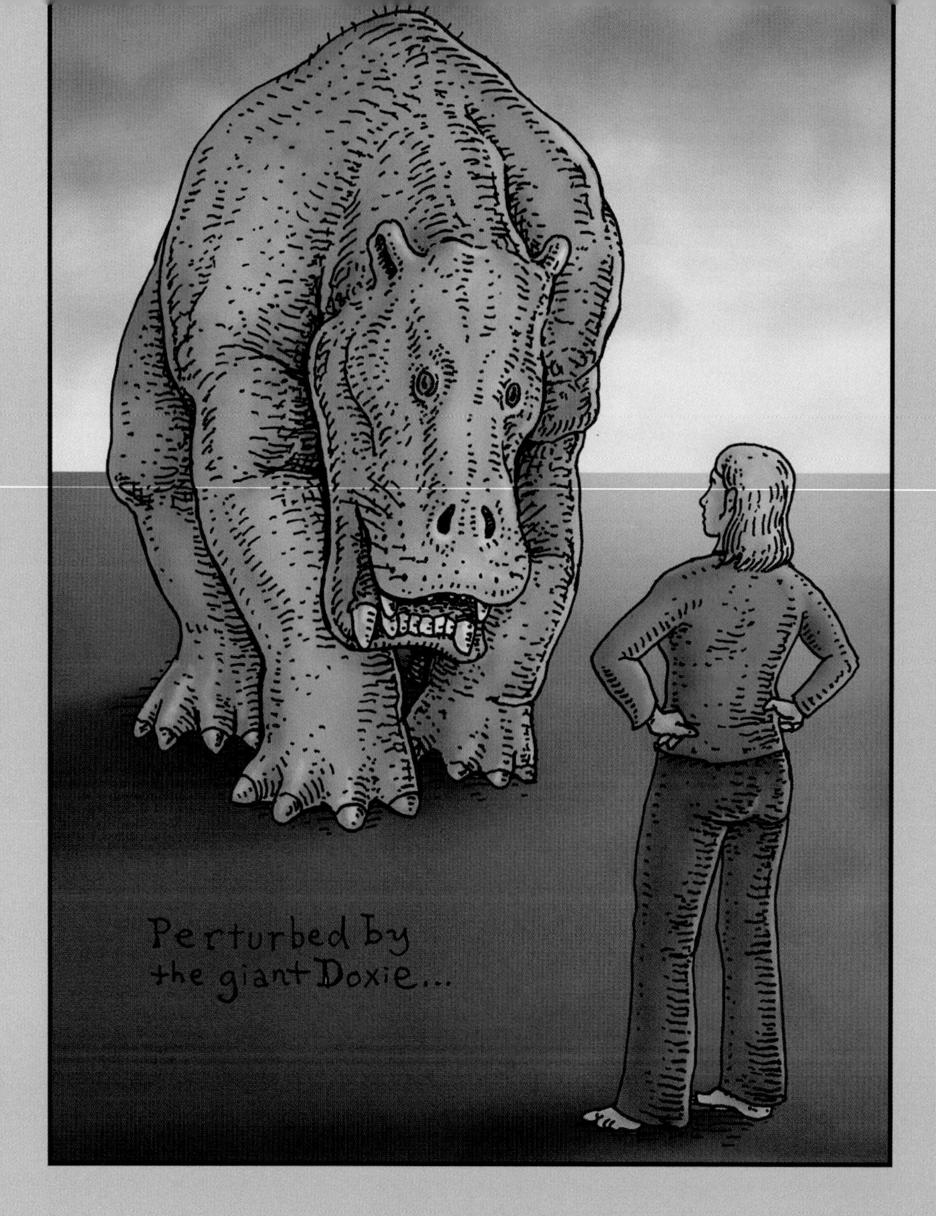
Perturbed by
the giant Doxie...

Orange County's Capistrano Formation. The skeleton had been discovered in 1993 when a bulldozer scraped through his left scapula – he had been curled up in a ball like he had been sleeping. The same formation has produced fossil leaves and flowers and lovely mantis shrimp.

Out behind the main building, the ground was littered with forty-four car-sized plaster-covered blocks from the Late Eocene Talega Formation near San Clemente. The blocks were chunks from a huge bone bed that contained rhinoceroses, brontotheres, camels, tapirs, snakes, turtles, and crocodiles. There was enough work for a team of fifty; Meredith had a few employees and a team of forty or so volunteers.

There was another building full of fossil whales from various sites around Orange County and then a couple of containers full of whale bones from a bone bed in Irvine. There were enough fossils here to stock a major museum. Back in the main building, Meredith opened the locked cabinets to reveal even more treasures. She had a splendid *Megalonyx* sloth skull from nearby Clark Park and the skull of *Phoberogale*, the earliest known bear in North America. There were splendid Cretaceous palm fronds and ammonites and a variety of Native American artifacts.

It had become clear that the past fifty years of economic growth in Orange County had led to a surge in fossil discovery and excavation in a place that had no museum to handle the influx. This county has a beautiful story to tell and the Cooper Center is a first step toward building that museum.

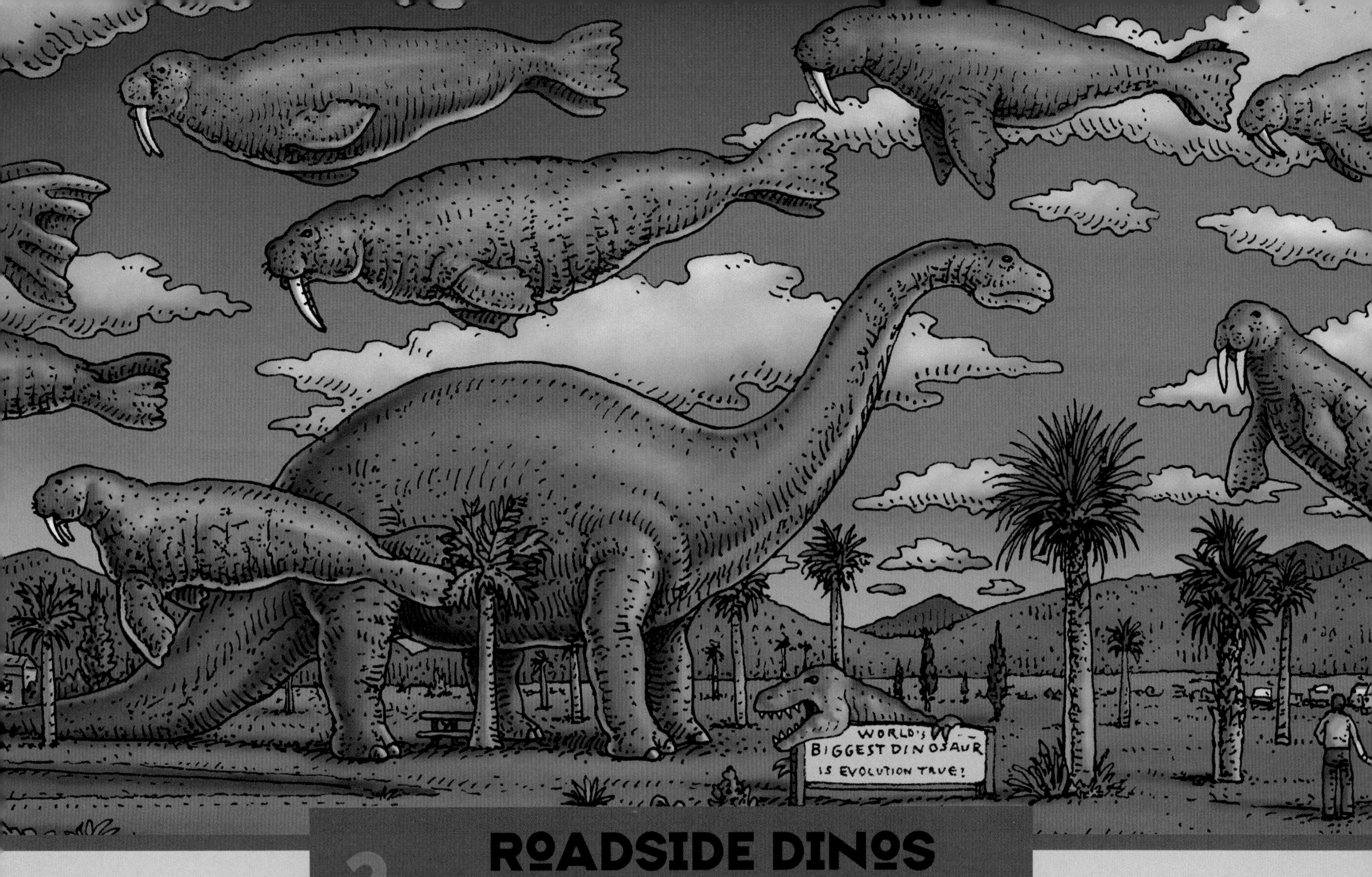


# 2 ROADSIDE DINOS AND DESERT WALRUS

I have a walrus problem, and it started with whales. In 1982, I took a job working for the Marine Geology Branch of the US Geological Survey in Menlo Park, California, and was given the task of mapping the feeding grounds of the California gray whale in the Bering Sea of northern Alaska. At the time, there were about 15,000 of these 40-foot-long mammals. They would winter in several different lagoons along the west coast of the Baja Peninsula and then swim along the entire length of the West Coast to the northern Bering Sea, where they would feed on small seafloor sand fleas called amphipods.

My job was to scan rolls of paper that had been printed out during ship surveys of the Bering Sea. The surveys deployed a sonar device that sent a radiating sound pulse to the seafloor. The machine printed out the returning signal, which was essentially an image of the seafloor created by a sound wave, and the images were clear enough

A surreal vision of Pliocene desert walruses sailing through the skies above "Dinny the Dinosaur" and "Mr. Rex."

that I could easily recognize rocks, sand ripples, and even anchor chains. The survey ships had taken crisscrossing paths to sample the entire area north of St. Lawrence Island and south of the Bering Strait.

Each day I would scroll through each roll and note where individual seafloor features occurred. When I compiled all of my notes, it became clear that the entire floor of the Bering Sea was pockmarked by shallow oval pits that were about 6 feet long and 3 feet wide.

Gray whales have mouths that are about 6 feet long, and it quickly became apparent that we were mapping a submarine whale feeding ground. It turned out that gray whales suck their food off the bottom of the muddy seafloor. Imagine if cows grazed by sucking sod and spitting out the dirt, and you have a good sense of how gray whales feed.

Every day, I would go to my office and map gray-whale seafloor suck marks. As I was doing this, I began to notice long, thin, wiggly lines on some of the sonar images that were taken in water depths of around 100 feet. These traces were about a foot wide, 300 to 400 feet long, and looked like trenches. I wondered what kind of process could make such long, meandering, shallow trenches in the seafloor. I called around and eventually found myself talking to Mary Nerini, a scientist at the National Marine Mammals Laboratory at Sand Point in Seattle. Mary had just led a cruise to the Bering Sea to study marine mammal–feeding ecology and had taken professional divers with her. On some of their dives they encountered trenches that were about a foot wide, a foot deep, and extended for great distances. They used underwater whiteboards to map the trenches, and, as they mapped, they noticed that every couple of feet, they would encounter an open, empty clamshell. Whatever was making these trenches appeared to be eating clams. I was really puzzled by a whale making such a narrow trench, and I didn't think whales ate clams.

Later that year, I visited the museum in Fairbanks, Alaska, to inspect Bering Sea sand fleas and clams. While I was there, I dropped in on Bud Fay, the world's expert on walrus, and asked him what he knew about their feeding patterns. Bud presented me with a copy of his freshly published monograph about the Pacific walrus and told me a story about the time he had been with some St. Lawrence Island Native Yupik people when they had killed and eaten a walrus. One of the first things they had done after dispatching the animal was to open its body cavity and extract its stomach. In the stomach were the bodies of hundreds of clams. The stomach acid of the walrus had pickled the clams, and the hunters ate them like the

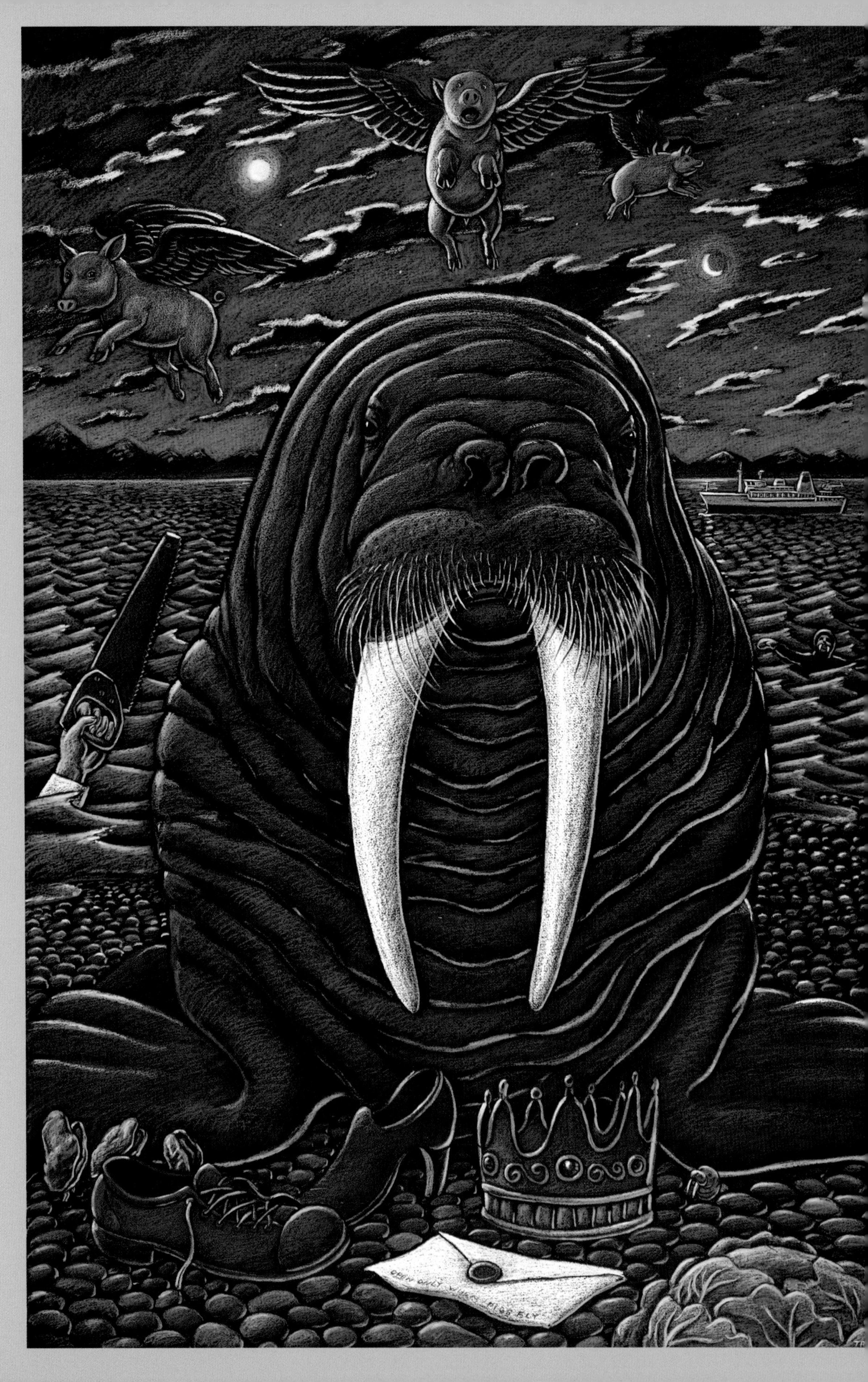

treasured hors d'oeuvres they were. Bud got curious and started counting the number of clams in the bellies of subsequent kills. One walrus whose stomach contents he counted in 1975 had an unbelievable 6,378 clam bodies in its stomach and no clam shells.

Above: Cretaceous shale on the shore at Point Loma near San Diego. Note the inoceramid clam fossil in the foreground.

Left: The walrus, with apologies to Lewis Carroll.

It was pretty clear to Bud that the walrus shucked clams before eating them, but it was not so clear how they got them out of the seafloor in the first place. He figured that it must have something to do with their long ivory tusks and their tactile whiskers, known as vibrissae. The ivory tusks of walrus can grow to 3 feet or longer and have been prized trading objects for centuries – many more people have seen walrus ivory than have ever seen an actual walrus. As obvious as the tusks are, their function is less clear. Wide-ranging ideas have included ice picks, seal stabbers, clam diggers, grappling hooks, and demonstrations of dominance.

When Bud inspected the tusks, he found wear patterns along the fronts rather than at the tips. That suggested to him that the tusks functioned more like sled runners than digging sticks. Likewise, the vibrissae were more worn on the front of the muzzle than on the sides.

When he looked at live walrus in captivity, he realized that their round tongues were a perfect fit for their rounded palates, and that a walrus tongue could work like a piston to jet water on the outstroke and create strong suction on the instroke. All of these observations came together in the idea that walrus likely fed by essentially standing on their heads on the seafloor and excavating clams by jetting water, finding them with their vibrissae, and shucking them by sucking the meat out of the shell. It wasn't until 2015 that *National Geographic* photographer Paul Nicklen was actually able to photograph feeding walrus in the Canadian Arctic and confirm this bizarre notion.

Walrus live mainly around the ice floes of the Arctic Ocean. Unless you are an Alaskan Native, an Arctic marine researcher, or a Bering Sea commercial fisherman, it is pretty unlikely that you will ever see a wild Pacific walrus. In 1985, while on a research cruise in the Bering Sea, I had the chance to visit Round Island in Alaska's Bristol Bay to see a summer walrus haul out. We arrived on a sunny day with flat, calm seas and were able to land on the island and climb across a steep snowfield and down to the top of a cliff that overlooked the beach full of walrus. It was a lucky day, and more that 10,000 closely packed male walrus lay on the rocky shore. The place smelled like a pig farm, and the sound of thousands of snorting, farting, tusk-clacking walrus is not one that I will ever forget. We sat and watched the walrus for hours until reluctantly taking our leave. I have always wanted to return.

None of this was on my mind when Ray and I walked into the San Diego Museum of Natural History at Balboa Park on a sunny day in May. I had never been to this museum before and had no idea what to expect. I did know that the paleontologists at the museum had been exploiting an interesting opportunity by hiring themselves out as salvage paleontologists for southern California construction projects. Not only did this approach bring much needed resources to the museum, it also gave a museum home to the fossils that were recovered during construction.

Within moments of arriving, it became very clear that this approach had resulted in a focusing of the museum's mission. By collecting local fossils, the museum had realized that it could tell the story of the planet's history through the lens of San Diego and the Baja Peninsula.

We arrived just a year or so after the museum had opened a new exhibition on local paleontology, and almost all of the fossils on display were ones they had recently found in and around San Diego. The open atrium space was full of skeletons and models of marine mammals, sharks,

Skeleton of the walrus *Valenictus chulavistensis* on display at the San Diego Natural History Museum

Kirk and Tom Deméré comparing male and female *Valenictus* walrus skulls in the collections of the San Diego Natural Museum History.

fish, turtles, and mammoths. The walls were plastered with compelling prehistoric murals painted by the artist William Stout. But what caught my immediate attention were the walrus fossils.

I had been aware that walrus were related to seals and sea lions and that they had evolved from short-toothed ancestors. I had even seen the short-tusked walrus named Waldo at the Cooper Center. What surprised me were the long-tusked fossil walrus from San Diego.

Right there on display was a nearly complete fossil walrus skeleton in clam-feeding posture. In the case next to it were the skulls of two different fossil walrus species. The first, *Valenictus chulavistensis*, looked at first glance like the skull of the living walrus, but on closer inspection, it was clear that the tusks were sharper and the skull more delicately built. The second, *Dusignathus seftoni*, looked more like a saber-toothed cat with medium-length dagger-like tusks. Both of these skulls had been collected by salvage paleontologists in the Chula Vista neighborhood of southern San Diego. This was something that I simply had not expected to see.

We had made an appointment to meet Tom Deméré, the head of the museum's paleontology program, and by this point I was really excited to talk with him. Tom welcomed us back to his well-lit and tidy office, where he told us about the success of the museum's salvage project and how he had shaped it into a full museum effort that resulted in great exhibits and more fossils. Along the way, he had honed his expertise in the paleontology of marine mammals and had become an expert in walrus evolution. Since I was a paleobotanist traveling with an artist, I don't think he expected the gush of walrus enthusiasm that was bubbling out of me as we sat down to talk.

Tom started working at the museum in 1979, just as the construction boom in San Diego was picking up steam. He quickly realized the virtuous cycle that could emerge. If the museum hired paleontologists to do salvage work, it would yield collections and science that would inform museum programs and exhibits, both of which needed money that could be supplied by salvage work. The paleontology team had grown to fourteen staff with five curators, making it one of the largest programs in the country.

It didn't hurt that the paleontology of San Diego was rich enough to support the effort. The native scrubby brushland of the region covered most of the natural outcrops, so it was only excavation for construction purposes that exposed the fossil-bearing rocks, and Tom's team was there waiting for the fossils when they came into view. One of the occupational hazards of this type of work is that fossils are often found by the blade of a bulldozer, resulting in many skeletons with bits sliced off. Of the new exhibit, 90 percent of the specimens were salvaged from construction sites, and about 50 percent have been sliced by bulldozers, à la Waldo at the Cooper Center. The haul has included fossils as diverse as whales, mammoths, and oyster-encrusted dinosaurs.

Tom Deméré and a cast of the skull of the prehistoric walrus *Valenictus chulavistensis*.

Fondling a plastic cast of a *Valenictus* skull, Tom gave us a quick lesson in walrus evolution. He was pretty sure that walrus had first evolved from other pinnipeds in low latitudes during the last 20 million years or so. Their origin predated the ice ages so their relationship to sea ice was something that happened late in their evolution. They appeared first in the Pacific, and by about 5 million years ago they started showing up in the Atlantic. I found it wholly bizarre that key members of the walrus family would not be known were it not for a San Diego subdivision.

And the Chula Vista subdivision was not just rich in walrus. In 2000, the team recovered the skeleton of a 3.5-million-year-old, 30-foot-long Pliocene sea cow named *Hydrodamalis cuestae*. This animal was related to the infamous Steller's sea cow, which was discovered by Vitus Bering to the west of the Aleutian Islands in 1789 and promptly eaten into extinction by his shipwrecked crew. Like frosting on a cake, the Pleistocene gravels of Chula Vista contain mammoth bones. This is a subdivision that should be inhabited by paleontologists.

In 2017, Tom's team published their findings from a San Diego highway salvage site that included fractured mastodon bones, teeth, and tusks in association with round rocks that were covered with percussion marks. To them, this implied that the bones were broken by humans. They dated the site at 130,000 years, which is roughly ten times older than the scientifically accepted evidence for the first humans in North America. Maybe some archaeologists should live in that subdivision as well!

Elsewhere around San Diego, the rocks range in age from a few thousands of years to more than 80 million, and many of them are rich in marine fossils. One crew working up the road near President Richard Nixon's home in San Clemente dug into Miocene Talega formations, and found the complete skeleton of a juvenile *Desmostylus*, two leatherback turtles, *Allodesmus* sea lions, fur seals, sea cows, river dolphins, and leaves.

The San Diego Formation, a late Pliocene deposit, yielded a veritable flock of marine birds such as grebes, boobies, gannets, auklets, flightless auks, and albatross (but surprisingly, only one gull). It also produced lots of fish, including stingrays. It is a property of nearshore marine deposits that land plants and animals are sometimes swept out to sea and are buried with their marine counterparts. Because of this, it is not unusual to find pinecones and whale bones in the same outcrop. So it isn't a huge surprise that the San Diego Formation yielded parts of a ground sloth and a young camel that had been bitten by a large shark.

The rocks near Carlsbad are Cretaceous and contain fantastic ammonites, cycads, palm fronds, and dinosaurs. All of these fossils and more on display in the museum were literally dug out of local sites and tell a very complete and utterly local story. Stunned by what we had seen, Ray and I drove down to Solana Beach for a walk in the surf. I immediately started finding fossil clams in the rocky cliffs. Ray struck up a conversation with the lifeguard and learned that a great white shark had killed a surfer here in 2008. We took a very quick dip.

That night, we stayed with a friend who threw a party for us. We met people who regularly ventured down the long Baja Peninsula. They told us stories of amazing fossils and incredible landscapes. We talked about the Sea of Cortez and the tiny porpoises called vaquitas (little cows), which were endangered because they were just the wrong size and ended up in the nets of anglers chasing totoaba, itself an endangered fish whose swim bladder has value in the Asian traditional medicine market. We knew that deserts were fabulous places to find fossils because plant cover was thin, and we realized that the Baja Peninsula must be a fossil paradise.

The next day, we headed east over the mountains toward another desert – specifically, the Anza-Borrego Desert, whose windswept badlands were full of fossils. Rolling down from the mountains, we could see the Salton Sea, a curious salty lake well below sea level that was formed by accident in 1905 when an irrigation ditch carrying Colorado River water broke out of its course and began dumping water into a depression that had been a lake during the Ice Age. The result was an unnatural lake that began to evaporate away the day that it formed. On a map of California, the Salton Sea appears to be the dribbled northern cousin of the Sea of Cortez. In actual fact, the Salton Sea was the northern extension of the Sea of Cortez until the delta of the Colorado River filled the Imperial Valley and cut the sea into two parts. Under the intense desert sun, the northern half evaporated away and stayed that way until the accident in 1905.

After a long downhill run, we emerged onto the wind-blasted desert floor and were startled to see a host of Ice Age animals running across the thorny landscape. There were mammoths,

Ricardo Breceda's steel sculptures of Ice Age mammoth, horses, and saber-toothed cats in the desert at Anzo Borrego.

horses, saber-toothed cats, and a bunch of dire wolves. On closer inspection, the animals were made of steel and the rusty sculptures were scattered across the landscape. We walked among them and could see no signs of who made them or why they were there. We later discovered that these sculptures were the work of artist Ricardo Breceda. The landowner of this patch of desert inaptly named Galleta Meadows had allowed his property transformed into a Pleistocene park for Breceda's sculptures.

The badlands to the west of the Salton Sea are formed of Pliocene and Pleistocene sediments, and they contain both expected fossils like mammoths and tortoises and completely surprising ones like walrus. We had come to see *Valenictus imperialis*, the imperial desert walrus, an animal that had lived in the Salton Sea when it was a northern extension of the Sea of Cortez. For a walrus lover used to sea ice, this was one of the most incongruous situations I had ever witnessed. We approached the visitor's center and museum in Borrego Springs with tremendous anticipation. What a phenomenal opportunity to juxtapose the mighty mammoth with a desert walrus to create a jarring educational opportunity that would open eyes to the incredible impact of time and climate on the distribution of animals.

It was also hot as hell and we were delighted to enter the cool building, prepared to enjoy an afternoon of interpretive bliss. Then… What? The center had almost no fossils, few murals, and miraculously, no mention of walrus at all! I queried the docent about the dearth of walrus. "What walrus?" he replied. Another pinniped interpretive moment squandered. We headed north to Palm Springs.

Off to our east, we could see the desert mountains of southeastern California. Most people see these ranges as they are driving to or from Las Vegas and think nothing of them. And yet a few months earlier, I had visited with a crew of MIT geology students, and we had climbed around on outcrops of early Cambrian rock and collected ollenelid trilobites. At 520 million years, these rocks actually house some of the very oldest animal fossils on the continent.

This concrete *Brontosaurus* is so large, it has a gift shop in its belly.

Claude Bell's giant *Tyrannosaurus rex* at Cabazon is a bit too large and incorrectly has three fingers instead of two.

We reached the interstate and began to arc back toward the Los Angeles Basin. Traffic began to increase. As we crossed a low pass, vast fields of wind turbines appeared, and the sunny haze of the evening began to dull the light.

I noticed that we were about to drive through the town of Cabazon. Most people know Cabazon from the 1985 movie *Pee Wee's Big Adventure.* I know cabezon as a giant north Pacific sculpin (*Scorpaenichthys marmoratus*) that eats crabs and thus tastes deliciously like crab. I know the town of Cabazon as the hometown of western historian Patty Limerick. But having never driven through Cabazon before, I didn't realize that it was the spot that a man named Claude Bell had chosen to make his mark. As we approached the town, a pair of huge gray dinosaurs loomed into the sunset sky.

In 1964, after retiring from a long career at Knott's Berry Farm, sixty-three-year-old Claude Bell stated that "there's lots to do before the sand runs out" and set about building a huge concrete *Brontosaurus* on his property in Cabazon. The effort took him eleven years, 50 tons of steel, and 100 tons of concrete. The mighty beast is 150 feet long, 45 feet tall at the hip, and large enough in the belly to host a large and well-stocked gift shop. After taking a five-year break and still full of energy, seventy-nine-year-old Claude got back to business and constructed a 55-foot-tall *Tyrannosaurus rex.* His wife Anne Marie said, "He's not off his rocker, it's not like he's just building some dumb thing out in the desert. He just wanted to do something that wouldn't be torn down.

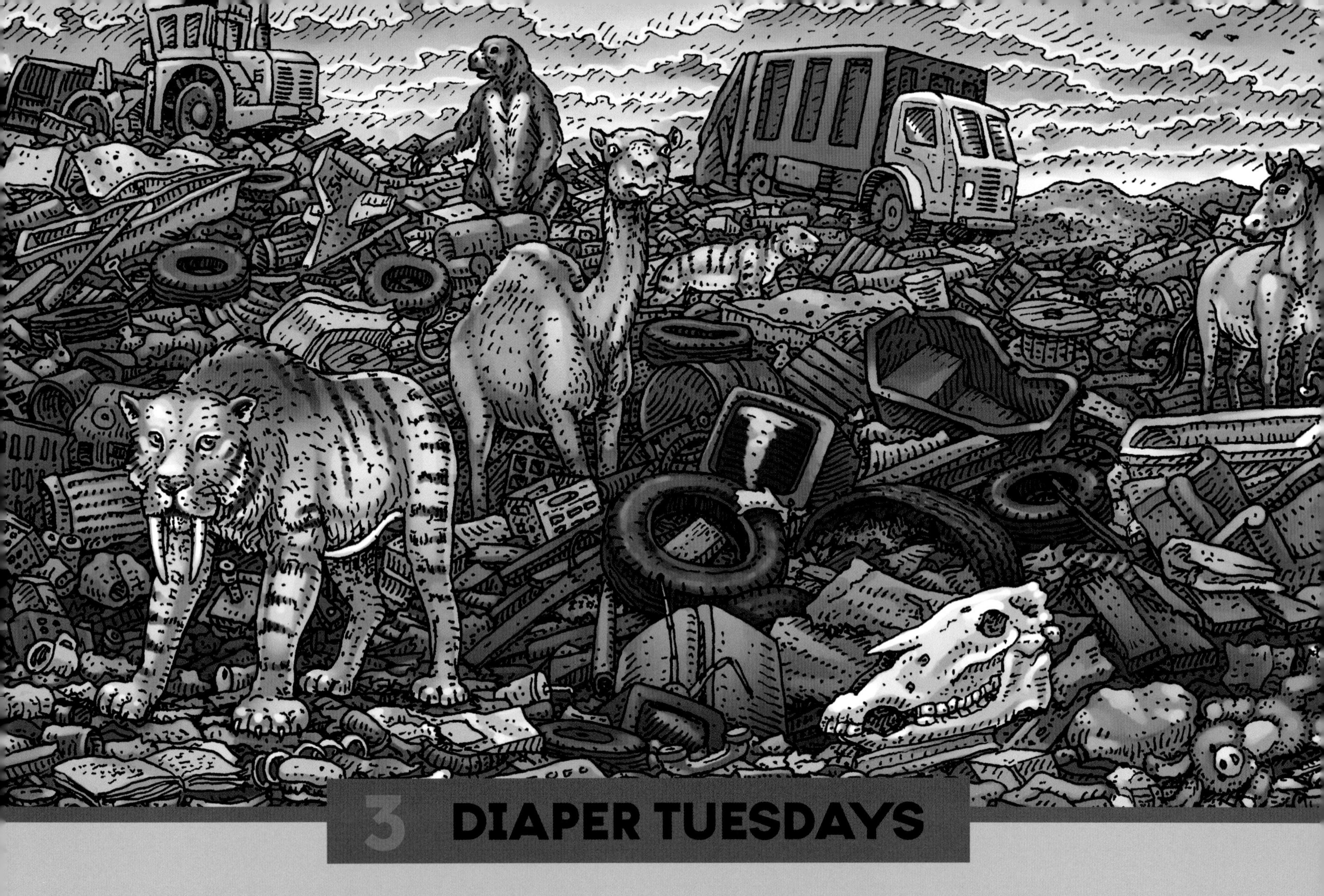

# 3 DIAPER TUESDAYS

The eastern side of the Los Angeles Basin is a place that has deep ties for me because my parents lived there in the 1950s. My dad attended Loma Linda Medical School, and my mom edited the medical school's alumni journal. They met and spawned a paleontologist. This is somewhat ironic because Loma Linda is a Seventh Day Adventist (SDA) school, and SDAs are the original American creationists. It was SDA George McCready Price who, in 1915, watched the advances of paleontology and geology and saw the writing on the wall for his biblical view of the world. He realized that the data emerging from the fossil fields was beginning to strongly support a Darwinian worldview, and he figured there must be another way to interpret those new findings. His writings were built on the insights of the emerging fossil record, and he conceived the idea that the flood of Noah was responsible for the stratigraphic record that we see today.

The title of one of Price's early books, *God's Two Books*, referred to the Bible and the geo-

An assortment of Ice Age mammals roaming a modern-day landfill near the town of Chowchilla, California.

logic record as dual references to the same set of events. In a time before plate tectonics, radiometric dating, and DNA, it was an argument that made sense to the congregation, and the SDAs became the leading proponents of creationism and a 6,000-year-old Earth. One curious by-product of this history is that the SDAs were unusually interested in stratigraphy and fossils. They had an axe to grind, and the stratigraphic record was one of their tools. As the century rolled on and science blossomed, however, their argument crumbled into powder. This fact was not noticed by the church, whose beliefs could not loosen to accommodate the new knowledge. So it was that I grew up attending a church that was dogmatically creationist but very interested in fossils and geologic time. In some way, my becoming a paleontologist was not a rebellion at all but simply a result of paying attention to the discussions at church.

As a result, I clearly have mixed feelings when I drive through Loma Linda. On campus, geologic time is still a touchy topic, and the geology department has to tread very carefully lest it offend the university's administration and the on-campus Geoscience Research Institute. A recent (2000) stratigraphy textbook by Leonard Brand, one of Loma Linda's geology professors, presents a markedly upside-down view of the rock record. His approach might have flown in 1860, but it is largely outside the realm of science today. I still have relatives and friends who live in Loma Linda and subscribe to the church's teachings. I'm actually grateful to the church for introducing the idea of evolution to me when I was a kid. By pushing the opposite view so hard, they accidentally gave me earlier and more effective introduction to science than I might otherwise have had.

But just because they drove me to paleontology doesn't mean that their message has always backfired. In fact, the percentage of Americans subscribing to a message of a young Earth and a week of creation has remained stubbornly high, even as the scientific evidence for Earth's antiquity continues to pour out of the ground. And ironically, a lot of that evidence is found not far from Loma Linda.

The eastern part of the Los Angeles Basin is full of fossils and paleontologists, and I had long admired the work done at the San Bernardino County Museum in Redlands. Like many paleontologists in southern California, Eric Scott and Kathleen Springer spent their careers salvaging amazing fossils from construction sites. In 2014, their work led to the naming of the Tule Springs National Monument to the west of Las Vegas. If you were to run into Kathleen or Eric on the street, you might think Eric was on his way to a business meeting and Kathleen was on her way to catch some waves; neither looks like your typical paleontologist.

The museum was closed when we showed up, and we wandered through the half-finished exhibit about local geology and paleontology. It was a fun exhibit, complete with a California geology kitchen where visitors could cook up their own stories about the local rocks and fossils. Down in the collection rooms, we got a chance to see some local treasures. And just twenty minutes from the museum, the road over Cajon Pass displays more than 70 million years of fossil-rich local history. From dolphins to horses and camels, the world-class fossils from the valley walls paint a dynamic picture of San Bernardino County.

Our real goal for visiting Kathleen and Eric was to scope out the Western Science Center, which opened in Hemet in 2006. The museum's website called itself the "Smithsonian of the West," and it had become a Smithsonian Affiliate museum in 2008. Since I was a newly minted Smithsonian museum director, it seemed worth a look. We hopped in Kathleen's Jeep and headed south.

The fossil saga of Hemet started back in 1993 when the Metropolitan Water District of Los Angeles decided it might be prudent to build a reservoir east of the San Andreas Fault. With 20 million people living to the west of the fault, it seemed like a good idea in case of a major earthquake. There were not a lot of good sites, so they settled on Hemet and began to build a curious reservoir between two east–west trending mountain ranges, damming both the east and the west drainages between the mountains with a pair of huge earthen dams. Once filled with Colorado River water (what else?), the double-dammed reservoir was initially called the Domenigoni Valley Reservoir, but it is now called Diamond Valley Lake. The giant bathtub was to become the largest reservoir in southern California, and it would have a capacity of 300,000 acre-feet of water. To pull this off, they would have to move 210 million cubic yards of dirt! Knowing that both archaeological artifacts and fossils might be found on the site, the

budget included an amazing $6 million for salvage paleontology, $30 million for salvage archaeology, and $30 million to support the idea that a paleontology/archaeology/water museum would be built alongside the new reservoir. Work began in 1993.

Kathleen Springer was hired as a project paleontologist and got right to work. The fossils came fast and furious, and they had to race the bulldozers as the project moved forward. The digging roared forward at twenty hours a day, six days a week. The first fossil found was part of a mastodon, and thousands more soon followed. The digging went on for a decade and when the dust had settled, the fossil salvage crew had located 2,646 fossil localities and collected more than 100,000 bones. The picture that emerged was of a world 40,000 years ago

Above: Eric Scott harassing Kathleen Springer with the skull of a Miocene camel from Cajon Pass.

Below: Kathleen Springer pondering an earlier version of herself cleaning the teeth of a mastodon skull from the Diamond Valley Reservoir near Hemet.

that was populated by mastodons, mammoths, horses, camels, bison, and three different kinds of ground sloths. In some ways, these animals were similar to those found at La Brea, but there were significant differences as well – one being that carnivores, while present, were quite rare. From this view, it is pretty easy to see how the sticky tar at La Brea resulted in the overrepresentation of carnivores.

By 2000, parts of 140 mastodons had been collected, and the paleontologists lobbied to rename the lake Mastodon Valley Reservoir. Despite the fact that no diamonds were found, they ultimately lost the naming battle, and Diamond Valley Lake was christened in 2002. The museum opened in 2006.

We pulled into the parking lot of the Western Center, and I saw a place with a lot of potential. The huge complex was open but almost completely deserted. A kind volunteer lady staffed the admissions desk, but she appeared to be the only soul in the entire museum. We wandered around in the cavernous interior and saw some of the bones from the excavation. The visit was a bit tough on Kathleen, who had collected so many of the fossils only to see them displayed where there was currently so little audience to see them and no scientists to study them. The volunteer let us into the cavernous fossil collection facility where the results of the excavation were stored on huge shelves. We had a quiet afternoon, pulling drawers and looking at amazing fossils. This place clearly had tremendous potential, and we left hoping that someone would realize the opportunity. Fortunately, in the years after our visit, this has happened, and the Western Science Center has become another active node of southern California fossil zeal.

The Raymond Alf Museum on the campus of the Webb Schools in Claremont.

Our next stop was a peculiar pair of high schools that give Hogwarts a run for its money. For years, I had been hearing about the Webb Schools east of Los Angeles, which trained their students to be professional paleontologists. Malcolm McKenna, the famous American Museum paleontologist who launched the museum's second wave of expeditions to the Gobi Desert of Mongolia, was an alumnus, as was Dan Fisher, Michigan's mastodon and mammoth guru. Another alumnus was David Webb, Florida's Ice Age animal expert, who popularized the great interchange of mammals that happened between North and South America after the formation of the Isthmus of Panama. Ray and I decided that it was time for us to show up at the Webb Schools in Claremont and see what all of the fuss was about. I called ahead and discovered that Don Lofgren, a friend of mine, was now the director of the campus's Raymond Alf Museum.

We drove into a lovely understated campus that was built around an octagonal building that turned out to be the museum. We were met by Don Lofgren and Ashley Hall, a delightfully fossil-obsessed woman whom we had met the previous week at the natural history museum in Los Angeles. Ashley had picked up the fossil bug at the age of five and would relentlessly pester her parents to drive her from their home in South Bend, Indiana, to the Field Museum in Chicago. "All I ever wanted for Christmas was a trip to the museum. I knew all of the genera and species." She was living the dream by working at both the LA museum and at the Raymond Alf Museum, commuting the entire width of the LA Basin to appease her love for fossils.

The museum was full of excellent fossils and photographs of decades of field trips. It was also a temple to the memory of the man named Raymond Alf, a sprinter who came to Los Angeles to run for the LA Track Club in 1929. Since sprinting is not a career, Alf got a job teaching science at the Webb School (at the time, a boys-only school). At some point, he learned that there were fossils near Barstow, California, and, in 1936, he went there with some students from the school. One of the students found a fossil skull, and Alf took it to Pasadena to meet the paleontologist Chester Stock, who identified it as a peccary and published a description of it the next year. Alf would later say that the peccary skull lit the "flames of enthusiasm" in him. He was clearly one of those obsessed and inspiring individuals who finds their voice while cajoling high school kids. He was soon organizing "Peccary Trips" where he formed boys into bands and marched them off into the deserts of California, Nevada, Arizona, and the Rocky Mountain West in search of fossils and manhood. These trips grew to be the stuff of legends, and, now coed, they continue to happen to this day. By the mid-1960s, the collections from these expeditions had grown so large that Alf convinced the school to build its own museum to house them. The cult of the peccary created a steady stream of southern Californian high school students who were more than ready

Ashley Hall proudly presents the cast of a skeleton of *Amphicyon*, a Miocene bear dog that roamed the West.

for careers in paleontology. It was *Good Will Hunting* meets *Jurassic Park*, and it brought high school students into regular contact with active scientists. Many of the students went on to become scientists in their own right, and the Webb Schools became a veritable engine for paleontology. Alf died in 1999 at the ripe old age of ninety-three, but the engine is still running and the school is still churning out peccary-obsessed paleontologists.

The museum we walked into was full of all sorts of fossils, photos, and supporters of Alf. We met an older gentleman named Dick Lynas, who was Webb School class of 1955 and a veteran of many Peccary Trips. He showed us rows of giant titanothere skulls that were named by Alf for the boys that found them, including one named Richard. We were given Ray Alf swag, including a splendid snow globe complete with Alf reaching for the sky.

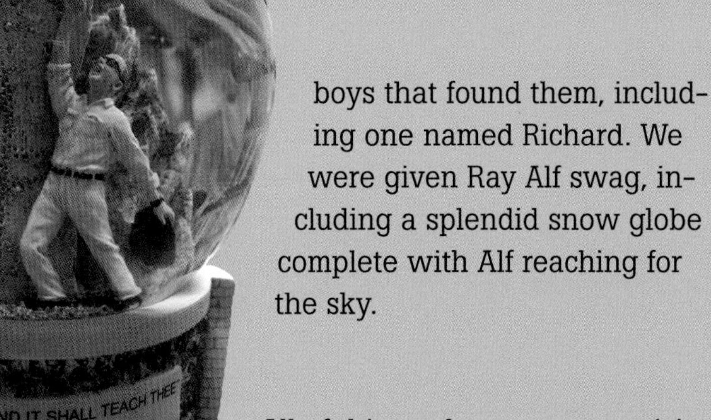


All of this made me want to visit Barstow to see the place that had lit Alf's flame. Barstow is a small town that lies about halfway between Los Angeles and Las Vegas. Historically, it is a place where trails, railroads, and highways crossed, so lots of people have been through Barstow but not so many have stopped. About 8 miles north of town lie the Rainbow Basin Badlands, an isolated patch of tilted and layered rocks that have been yielding a steady stream of fossils since the 1930s.

Beginning in the 1870s, paleontologists began finding more and more fossil sites across western North America, and it rapidly became clear that wave after wave of mammal species had lived on the continent. Really good fossil sites became exemplars of certain time periods, and the Rainbow Basin was one such site. Its rocks were deposited just about 13.8 million years ago, and the assemblage of animals from this time period was given the name "Barstovian" after the fossils found near Barstow. These time period names for groups of mammal species are known as "North American Land Mammal Ages" or "NALMAs." Now you know some real paleontological jargon!

Above: Ray Alf's quest, forever memorialized in a snow globe.

Left: Dick Lynas and the titanothere skull he found as a teenager.

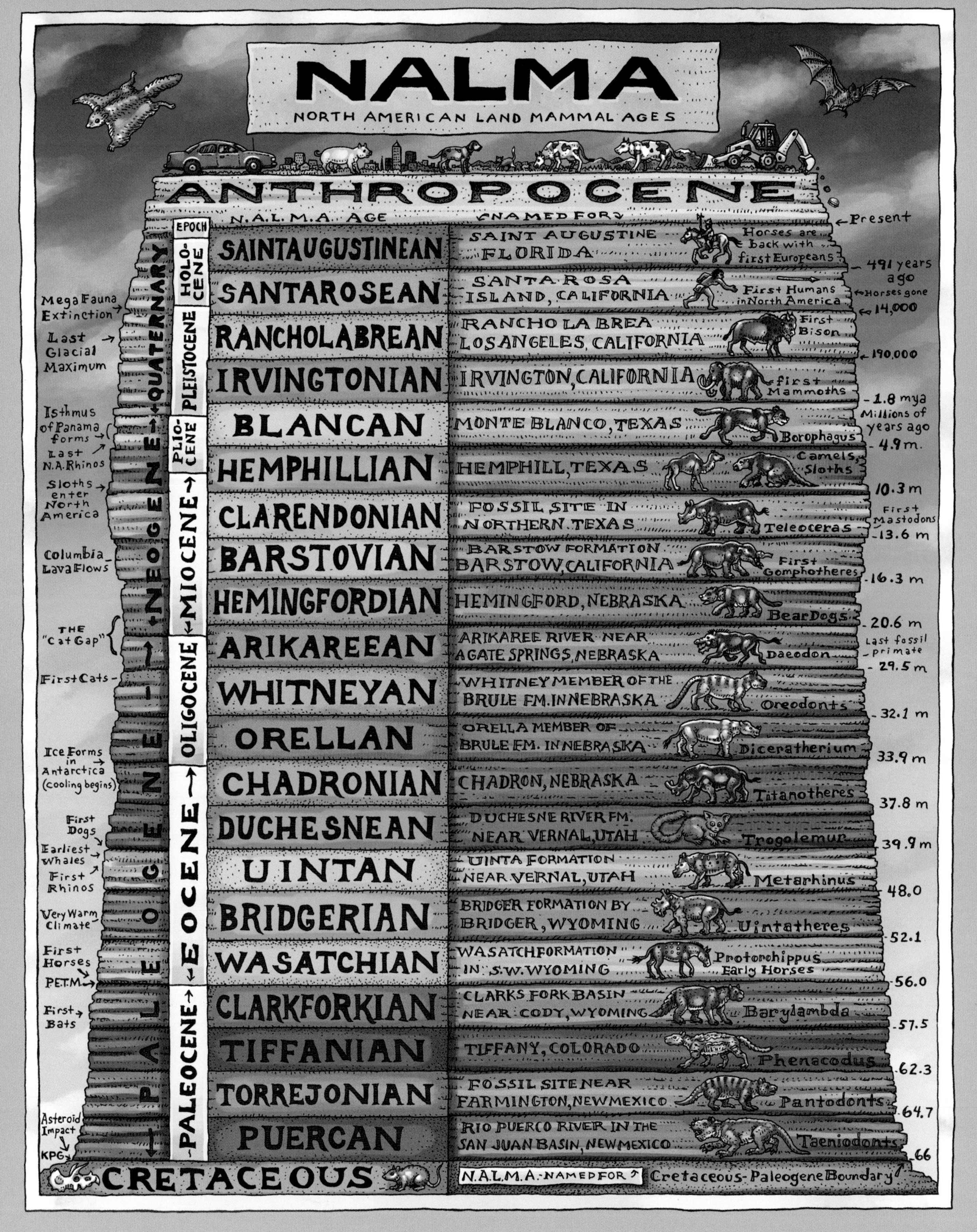

NALMA
NORTH AMERICAN LAND MAMMAL AGES
ANTHROPOCENE
EPOCH
N.A.L.M.A. AGE
NAMED FOR
Present
SAINTAUGUSTINEAN
SAINT AUGUSTINE FLORIDA
Horses are back with first Europeans
491 years ago
SANTAROSEAN
SANTA ROSA ISLAND, CALIFORNIA
First Humans in North America
Horses gone
14,000
RANCHOLABREAN
RANCHO LA BREA LOS ANGELES, CALIFORNIA
First Bison
190,000
IRVINGTONIAN
IRVINGTON, CALIFORNIA
first Mammoths
1.8 mya Millions of years ago
BLANCAN
MONTE BLANCO, TEXAS
Borophagus
4.9 m.
HEMPHILLIAN
HEMPHILL, TEXAS
Camels, Sloths
10.3 m
CLARENDONIAN
FOSSIL SITE IN NORTHERN TEXAS
Teleoceras
First Mastodons
13.6 m
BARSTOVIAN
BARSTOW FORMATION BARSTOW, CALIFORNIA
First Gomphotheres
16.3 m
HEMINGFORDIAN
HEMINGFORD, NEBRASKA
Bear Dogs
20.6 m
ARIKAREEAN
ARIKAREE RIVER NEAR AGATE SPRINGS, NEBRASKA
Daeodon
Last fossil primate
29.5 m
WHITNEYAN
WHITNEY MEMBER OF THE BRULE FM. IN NEBRASKA
Oreodonts
32.1 m
ORELLAN
ORELLA MEMBER OF BRULE FM. IN NEBRASKA
Diceratherium
33.9 m
CHADRONIAN
CHADRON, NEBRASKA
Titanotheres
37.8 m
DUCHESNEAN
DUCHESNE RIVER FM. NEAR VERNAL, UTAH
Trogolemur
39.9 m
UINTAN
UINTA FORMATION NEAR VERNAL, UTAH
Metarhinus
48.0
BRIDGERIAN
BRIDGER FORMATION BY BRIDGER, WYOMING
Uintatheres
52.1
WASATCHIAN
WASATCH FORMATION IN S.W. WYOMING
Protorohippus Early Horses
56.0
CLARKFORKIAN
CLARKS FORK BASIN NEAR CODY, WYOMING
Barylambda
57.5
TIFFANIAN
TIFFANY, COLORADO
Phenacodus
62.3
TORREJONIAN
FOSSIL SITE NEAR FARMINGTON, NEW MEXICO
Pantodonts
64.7
PUERCAN
RIO PUERCO RIVER IN THE SAN JUAN BASIN, NEW MEXICO
Taeniodonts
66
CRETACEOUS
N.A.L.M.A.-NAMED FOR
Cretaceous-Paleogene Boundary
HOLOCENE
PLEISTOCENE
PLIOCENE
MIOCENE
OLIGOCENE
EOCENE
PALEOCENE
QUATERNARY
NEOGENE
PALEOGENE
Mega Fauna Extinction
Last Glacial Maximum
Isthmus of Panama forms
Last N.A. Rhinos
Sloths enter North America
Columbia Lava Flows
THE "Cat Gap"
First Cats
Ice Forms in Antarctica (cooling begins)
First Dogs
Earliest Whales
First Rhinos
Very Warm Climate
First Horses
PETM
First Bats
Asteroid Impact
KPG

In addition to the much-admired peccaries, the Barstovian world also hosted horses, camels, oreodonts, bear dogs, dogs, cats, and rhinos. The Barstovian mammal that really catches my attention, however, is the gomphothere. This was a weird animal related to elephants and with tusks like an elephant. It's just that it had four tusks, not two. Two tusks projected straight out of the end of their lower jaw, while two more emerged at wider angles from the skull. The animals were big, but not as big as elephants. Their earliest occurrence anywhere in North America is here in Barstow, and the prevailing wisdom is that they wandered onto the continent along an earlier version of the Bering Land Bridge and quickly colonized the New World like so many species before and after.

The Australian mammologist Tim Flannery once spent a year in Boston studying the evolution of the North American mammal faunas. With his fresh eyes, he came up with a great analogy that he published in his book *The Eternal Frontier*. He argues that North America is like a room with three doors, and that mammal species enter and depart the continent through these portals. The first door was the Bering Land Bridge, a strip of land that connected Siberia and Alaska when the sea level was low. The second was the Isthmus of

The tilted layers of mudstone and sandstone in the Rainbow Basin near Barstow hold evidence of the first elephants in North America.

Panama, and the third was the land connection that linked Greenland and Scandinavia before the North Atlantic widened and severed it.

The Bering Land Bridge, often called Beringia, appears to have been intermittently open for something like 100 million years. The Panama Isthmus formed as South America drifted into contact with the lower part of Central America approximately 3 million years ago. The connection to Scandinavia is just the opposite, as it was broken rather than formed by the drifting of the continents. It was open about 60 million years ago but stopped being functional around 30 million years ago.

These three connections are known not only by their actual geology but also by fossil sites that tell the story of which species could travel between continents and when they traveled. Because fossils are only found in places where sediment was deposited, little patches of fossil-rich land take on unusual significance. It is for this reason that the first known gomphotheres in North America are from the Barstow area, and from that information we deduce that the Bering Land Bridge was open for business around 13.8 million years ago.

The paleontologists who figure this out have to find and excavate the fossils; figure out the sequence of rocks they occur in; prepare, illustrate, and describe the fossils; and roll all of this information into a narrative that is in constant revision. To complicate matters even more, the scientists are specialists who group themselves by the fossils they study. The plant, shell, and mammal paleontologists all go to different conferences, use different terminology, and have different priorities. Yet somehow they have begun to stitch together the complicated history of California.

The glue that holds paleontologists together are the papers they publish and the museums full of fossils. With these two critical elements, the thread of the story can persist and grow while careers come and go.

And over time, specific sites have continued to yield fossils, bringing back generations of paleontologists. The Rainbow Basin of Barstow was one of these sites; another was Sharktooth Hill near the dusty town of Oildale, not far from Bakersfield where Larry Barnes from the NHM had found his bone bed. As a West Coast fossil kid, I grew up hearing about a mythical place near Bakersfield where there was a hill full of giant shark teeth. My grandparents lived in Fresno, and they had friends who were rock hounds who had a lot of shark teeth. I wanted those teeth badly.

A shark tooth from Sharktooth Hill near Bakersfield.

In 1977, my grandfather Elmer passed away; my grandmother Flo followed him the next year.

1
2
3
23
22
25
24
26
2
2
4
21
5
20
27
6
19
18
7
10
13
9
11
14
15
17
8
12
16

## The Marine Animal Fossils Found at Sharktooth Hill

1. *Oncorhynchus rastrosus* – "spike-toothed" salmon
2. *Valenictus imperialensis* – Imperial Desert walrus
3. *Zarhinocetus errabundus* – long-snouted dolphin
4. *Peripolocetus vexillifer* – early right whale
5. *Carcharocles megalodon* – giant shark
6. *Mola mola* – ocean sunfish
7. *Odontaspis* sp. – sand tiger shark
8. *Imagotaria downsi* – early sea lion–like walrus
9. *Pelagiarctos thomasi* – bear walrus
10. *Aulophyseter morricei* – early sperm whale
11. *Allodesmus kernensis* – extinct pseudo-sea lion
12. *Semicossyphus pulcher* – sheepshead wrasse
13. *Neoparadoxia cecilialina* – "doxie"
14. *Desmostylus hesperus* – "desmo"
15. *Denebola brachycephala* – early beluga whale
16. *Psephophorus californiensis* - giant leatherback turtle
17. *Gomphotaria pugnax* – double-tusked walrus
18. *Dusignathus seftoni* – double-tusked walrus
19. *Atocetus nasalis* – ancestral delphinoid dolphin
20. *Parapontoporia sternbergi* – extinct "river" dolphin
21. *Balaenoptera bertae* – extinct minke whale
22. *Eschrichtius* sp. – gray whale
23. *Makaira nigricans* – extinct marlin
24. *Hydrodamalis cuestae* – giant sea cow
25. *Protoglobicephala mexicana* – early pilot whale
26. *Semirostrum ceruttii* – half-beaked porpoise
27. *Megachasma* sp. – megamouth shark

Elmer had been a gardener, so when my mother and I flew to Fresno in 1978 to clean out the family home, I found myself with an amazing inheritance that included a 1957 Chevy pickup truck and a garage full of digging tools. For a kid who hoped to someday become a paleontologist, this was an unexpected and extravagant bonanza. We spent the week sorting and selling items, and then I asked my mom for a favor. It is only 110 miles from Fresno to Bakersfield, and I figured that my mom and I had everything we needed to make an assault on Sharktooth Hill. Being the mom that she was, she agreed and we headed south.

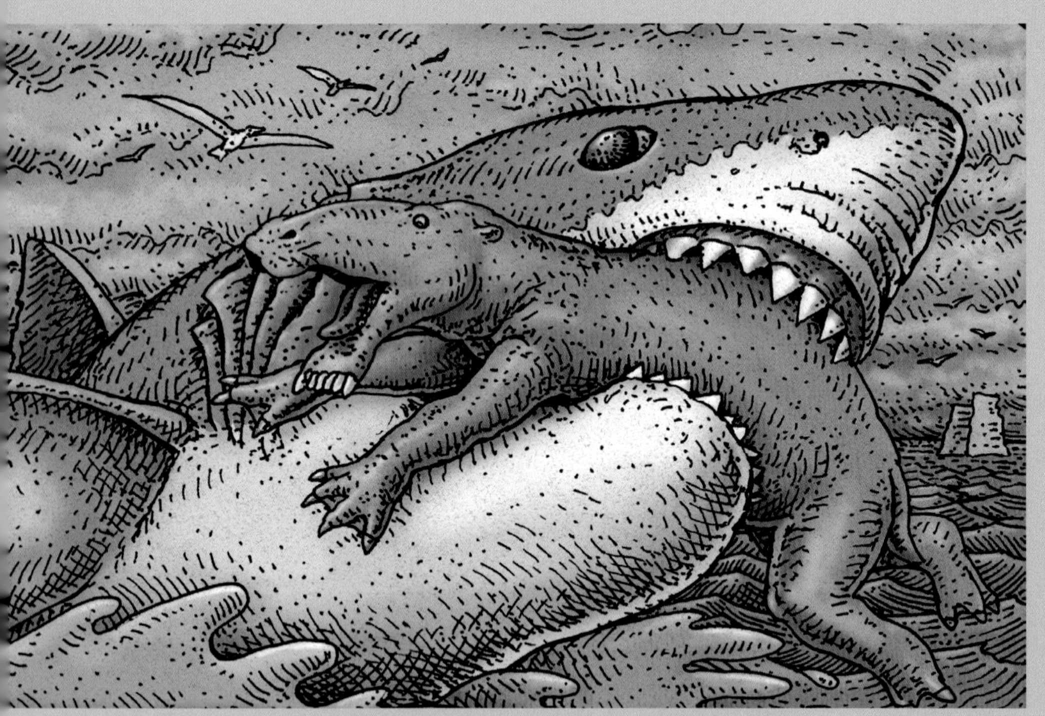

Looking back on it, I am amazed that I had enough information to get lucky. But get lucky I did. We drove to Oildale and up into a series of dusty hills looking for holes in the ground, and we immediately found them. I could not believe my luck; the holes were full of broken fossil bones and shark teeth. While my mom hung out in the truck and read magazines, I conducted my own personal excavation. At the time, I did not consider the fact that I was more than a hundred miles from the Pacific Ocean or, for that matter, whose property I was on. I was just plain giddy that I had a pickup truck and I could drive it across the landscape and find amazing fossils.

What I did not know at the time was that the bone beds of Oildale are some of the richest marine mammal sites in the world. They are so prolific that they were found soon after the California Gold Rush. The famous Harvard fish paleontologist Louis Agassiz was publishing on these fossils by the 1850s, well before the discovery of the fossil fields of the Rocky Mountains. Another famous fish paleontologist and founding president of Stanford University David Starr Jordan did research on the Sharktooth Hill sharks in the 1920s.

Sharktooth Hill is composed of sand deposited by a huge, shallow sea that filled what is now the Central Valley of California some 15 to 16 million years ago, and the bone beds in Oildale hold millions of bones of just about everything that lived in the sea as well as the animals that lived along its shores. The census of ancient marine life is massive and includes whales, dolphins, pinnipeds, sharks, other marine fish, seabirds, marine turtles, and our beloved desmostylians. In addition to the marine mammals, the bone bed contains the remains of land mammals that washed out to sea, including animals as diverse as true cats, bear dogs, horses, tapirs, rhinoceroses, camels, and gomphotheres.

# THE WHALE FAMILY TREE

By the 1930s, this was a major destination for paleontologists who were trying to understand the evolution of marine mammals. One of these was Remington Kellogg, an Iowa boy who studied paleontology at Berkeley before coming to the Smithsonian in 1928. In 1948, he became the director of the National Museum of Natural History (the position I now hold) and went on to found the International Whaling Commission. His early papers describe fossils from Sharktooth Hill.

Larry Barnes made his first forays to Sharktooth Hill in the 1950s, and he has been studying the fossils ever since. In 1983, he started removing layers of overburden with bulldozers to expose the bone bed and study how it came to be. After several studies, it became clear that rather than being formed by some mass stranding, the layer of bones represented a time period when strong currents washed sediment along the seafloor rather than letting it accumulate, so the only things that piled up were the carcasses of animals that died in the water column. A long period of no sand resulted in a dense layer of bones.

Many other paleontologists have studied these rich fossil beds, but one of them was a young amateur from Lincoln City, Oregon, named Doug Emlong. On April 15, 1975, Emlong found a complete pinniped skeleton at a place called Pyramid Hill. The site was a bit older than the main bone bed, and the fossil, now named *Enaliarctos mealsi*, is the oldest known pinniped in the world. When I discovered this fact, I realized that while I was digging a random hole in 1978, I was not far from where Doug Emlong had been making significant discoveries just three years earlier.

The site continues to be excavated and studied by scientists, dug by amateurs, and mined by commercial fossil diggers. One of the latter, Bob Ernst, actually bought parcels of land so that he could mine marine mammal fossils. A lot of his fossils ended up in the Buena Vista Museum of Natural History in Bakersfield, which offers summer digs for kids older than twelve during the summer.

A scimitar cat (*Homotherium*) and dire wolf lower jaws from the Fairmead Landfill in Chowchilla.

Ray and I drove north from Bakersfield into the broad expanse of California's Central Valley. What had been a sea during the Miocene had eventually filled in to become a vast plain. Early explorers to California's Central Valley pronounced it a wildlife paradise, describing endless flocks of migratory birds and vast herds of elk, antelope, and deer. They also talked about the ferocious California grizzly bear as a danger of the open grassland. When my great-grandparents arrived in Fresno from Nebraska in 1905, the valley was well on its way to becoming America's most productive agricultural landscape, and they set about growing grapefruit, peaches, raisins, and almonds.

Ray and I went to Fresno to meet up with Bob Dundas, who was studying a newly discovered fossil site in Fairmead, and we met him and his friend Jim Chatters at his office at Cal State–Fresno. Back in May of 1993, the nearby town of Chowchilla had started excavating a giant pit to serve as an extension of the Madera County Fairmead Landfill. Workers were surprised when a mammoth tusk appeared more than 30 feet below the land surface, and they were forced to bring in mitigation paleontologists to monitor the site as they dug. It turned out to be unusually rich, and the monitoring continues to this day. A small museum has even sprung up next to it.

After a few decades of research, it is clear that the site is about 780,000 years old, nearly twenty times older than La Brea. Fossils from this time period are known as Irvingtonian after a site discovered in Irvington, California, in the 1930s. This was a time before bison had moved over the Bering Land Bridge into North America, so while there were many of the same animals found at La Brea and Hemet, there were no bison. We spent a few hours inspecting fossils from the site that Bob had in his office. He had three different kinds of ground sloths – *Megalonyx*, *Paramylodon*, and *Nothrotheriops* – and four different types of carnivorous cats – the saber-toothed cat, the scimitar cat (*Homotherium*), a panther-like cat, and a cheetah. In addition to these were fossils from the short-faced bear and the dire wolf (*Canis dirus*). This was starting to look like a very dangerous dump.

Three guys and three ground sloths: Kirk, Jim Chatters, and Bob Dundas holding *Megalonyx*, *Paramylodon*, and *Nothrotheriops*.

Bob had to teach a class, but Jim had time on his hands so he agreed to accompany us to the dump. As we drove north, I put two and two together and realized where I had heard the name Jim Chatters before. He confirmed that he had been the archaeologist who collected an 8,900-year-old human skeleton on the banks of the Columbia River in Washington in 1996. The skeleton, which was one of the most complete ancient humans ever found in North America, came to be known as Kennewick Man. The discovery quickly grew controversial, as the local Umatilla people claimed the skeleton as theirs and anthropologists clamored to study the remarkable find. Chatters had become embroiled in the politics when he suggested that the skull did not look like a Native American but might instead have been from as far away as Siberia. It took nineteen years, lots of legal wrangling, and a whole lot of science before the analysis of ancient DNA did make the connection between Kennewick Man and Washington's Colville Indians.

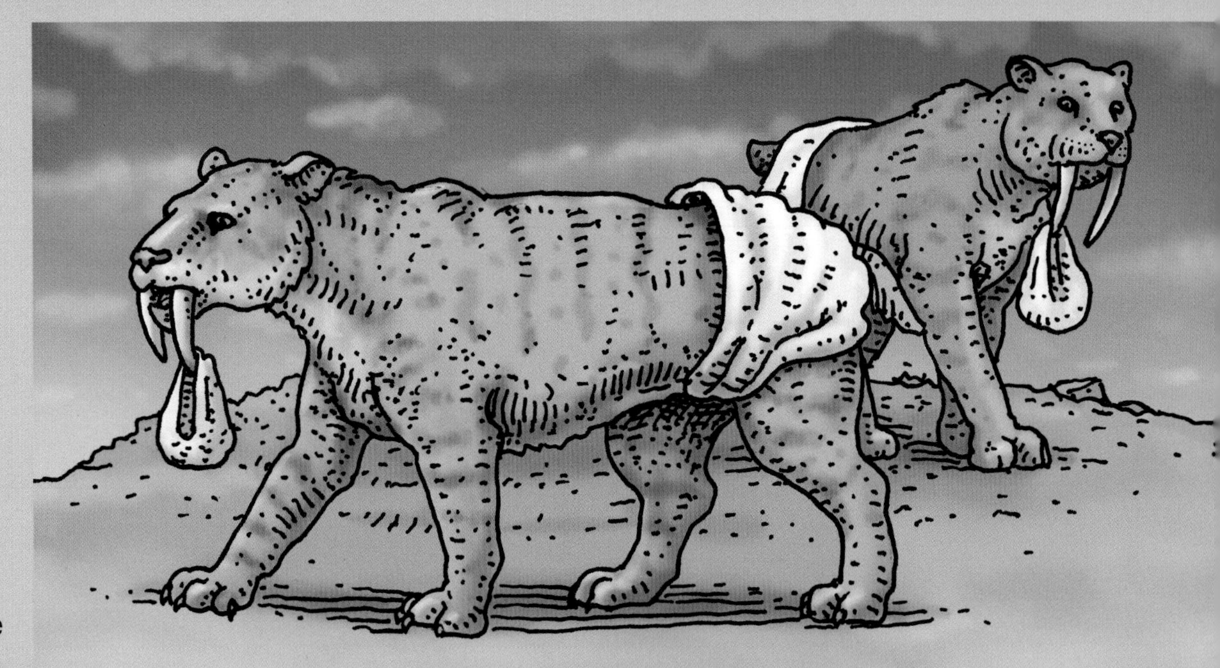

Ray peppered Jim with questions as we drove to Chowchilla, and it was clear that the Kennewick discovery was still a pretty sensitive topic. Since Jim had worked extensively in Washington, we shifted the conversation to other things, and Ray learned that Jim was a big fan of ratfish. Ray sometimes calls himself Ratfish Ray, so the conversational unease had cleared by the time we arrived at a city building in Chowchilla where the fossils from the dump dig were being stored.

Jim had been curating the thousands of specimens from the landfill and showed us around the old office building that had become a fossil repository. The bones from the dump were fractured, white, and crumbly, but they were definitely good enough to display the biological diversity of the ancient Central Valley. Despite the long list of carnivores, the most common fossils were herbivores, and the single most common herbivore was a species of horse that accounted for 61 percent of all the fossils (camels were 21 percent and sloths were 7 percent). Other interesting plant eaters included mammoths, a tiny antelope (*Caprimyrex*), a huge camel (*Camelops*), and an elk-sized llama (*Hemiauchenia*).

We were eager to see the actual site and arrived at the Fossil Discovery Center of Madera County to be greeted by a sour, penetrating odor that I don't usually associate with fossils. It was Tuesday, and Jim said that Tuesdays were always a bad day to be at the dump. I couldn't place the smell, but Jim referred to it as Diaper Tuesday and that made sense. It was clear that they had located the museum a little too close to the dump. The site was out in the center of the broad Central Valley and there was no tar here. The excavation in the dump itself showed that the fossils were buried in a deposit that was simply a layer of sand that had washed into the valley in a drier time.

The museum was full of tourists captivated by skeletons of the short-faced bear, ground sloth, and giant camel. A mammoth skull was hung from the rafters like a giant, bizarre ceiling fan. These were the early days for a museum that had only opened in 2010, and we understand that the floating mammoth skull now has a skeleton underneath it.

We left Jim at the museum and headed toward Berkeley, where we had agreed to do a book-signing event at seven that night. By four thirty, we were driving through oak-covered hills near Danville in Contra Costa County, and I remembered an old fossil story that connected the Seattle Seahawks with shovel-tusked elephants. I promised Ray that we had time to squeeze in one more fossil stop and still make it to the bookstore on time. He didn't believe me – and he was right – but we stopped anyway.

The floating mammoth skull at the Fossil Discovery Center of Madera County in Chowchilla.

A little background: A number of years ago while bartending for a party at the Explorer's Club in New York promoting the reintroduction of wolves into Yellowstone, I met a woman who told me that her brother, Ted Daeschler, was an environmental paleontologist. Even though I was a paleontologist, I had never heard of that title, and it baffled me until I finally ran into Ted in Philadelphia and learned that he studied paleoenvironments. That made more sense to me, and I learned that he had earned a master's degree at the University of California at Berkeley by studying a 9-million-year-old fossil site in California:

Ray hikes up a muddy road in Contra Costa County in search of the Blackhawk Quarry.

the Black Hawk Ranch Quarry. This was our pre-book-signing destination.

The Black Hawk Ranch Quarry is located on the slopes of Mount Diablo in the eastern San Francisco Bay area, and we just happened to be driving down Black Hawk Road when I had remembered the story about Ted. So, with some quick sleuthing on my iPhone, I was able to predict where we might find the quarry. It had just stopped raining, the sun had come back out, and the afternoon light was brilliant. By my calculation, we had to hike about a half mile up a steep dirt road to get to the spot where I hoped to find the quarry.

The road was soaked from the rain, and soon our shoes were coated in slippery mud that built up until we were slogging along on top of giant mud balls. Ray was skeptical, but I was pretty sure that we were headed in the right direction. Vindication came after about thirty minutes, when the road intersected with a very distinct human-made slash in the hill. The cut was only about 40 feet long and 10 feet deep, but it bore the clear signs of having been dug. The rock layers were tipped such that they were nearly vertical, and after a few minutes of crawling, we found a small piece of fossil bone. It wasn't much, but it was evidence that we were in the right spot.

Fossils were first found here in 1926 when the property was owned by the uncle of legendary landscape photographer Ansel Adams. Then, in 1937, the task of digging the site was assigned to the curiously named King Arthur Richey. It was during the Depression, and several Works Progress Administration workers were assigned to the project. It wasn't long before King Arthur and his diggers hit pay dirt and began extracting amazing plants and animals from the small quarry. Bones, teeth, and skulls emerged, and the list grew to include foxes, rabbits, beavers, pond turtles, deer-sized horses (*Hipparion*), camels (*Pro-camelus*), two kinds of lla-mas (*Pliauchenia* and *Megatylopus*), a ring-tailed cat (*Bassariscus*), a cougar-like cat (*Pseudaelurus*), a bone-crushing dog (*Epicyon*), a hyena-like dog (*Aelurodon*), and a large number of those shovel-tusked, elephant-like gomphotheres (*Gomphotherium simpsoni*). Later, as we researched the site, we learned that it had also produced the bones of *Oncorhynchus rastrosus*, the giant spike-toothed salmon. And while it is unusual for plants and animals to be preserved in the same quarry, this place also had leaves of poplars, elms, willows, sycamores, oaks, and sumacs.

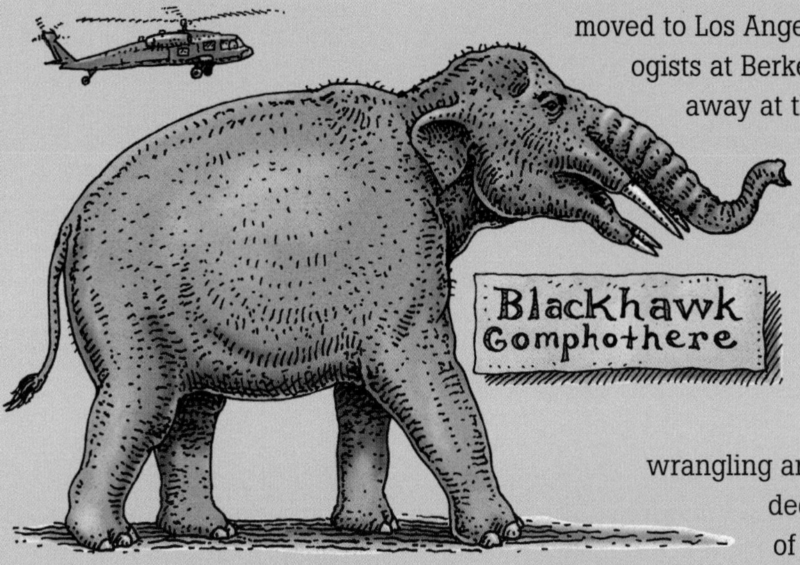

Blackhawk Gomphothere

King Arthur had excavated an entire Miocene world, and it was one that had formed just after the sea that once filled the Central Valley had drained. It was the first view of California as we know it, except that it had gomphotheres and giant salmon. Because of its rich fauna and flora, this was one of the first sites to be reconstructed as an ancient environment, a task taken by the artist William Gordon Huff, whose views of this world were exhibited in 1939. In 1941, King Arthur moved to Los Angeles, but the paleontologists at Berkeley continued picking away at the site for decades.

In 1975, Ken Behring, a developer and future owner of the Seattle Seahawks, purchased the property with the intent to build 4,800 homes. After much wrangling and local opposition, he decided to reduce the size of the development to 2,400 homes and to gift the quarry to the University of California. Ken was an eclectic collector of wild game mounts and rare cars. In 1991, he opened an automotive/natural history museum in Black Hawk. For its first six years, the museum had an exhibit of the Black Hawk quarry's fossils, and it was possible to see gomphotheres in amongst a hundred vintage cars. In 2003, he funded the Kenneth E. Behring Hall of Mammals at the National Museum of Natural History at the Smithsonian.

There was a lot of history in that small quarry on the side of the hill, and we ended up being late for our own book signing.

# 4 SURFIN' WITH THE DESMOS

There's a funny thing about natural history museums on the West Coast. For a place with so much natural history and so many people, museums are few and far between. This is one of the reasons why the fossils of the West Coast are not so well known to residents. Ray and I knew that the way to tell our story was to find experts – both scientists and local collectors – who knew what we wanted to know and to interview them. That modus operandi directed us both to local universities and to the towns where the best fossils were.

We knew that the University of California Museum of Paleontology (UCMP) in Berkeley would be a gold mine for our story and would consume several days of our visit. While most people think museums are places that only exhibit stuff, the reality is more complicated. They're also places where our culture preserves its treasures and where those treasures are studied. UCMP in

If the desmostylians of the North Pacific had survived to this day, one might have enjoyed a scene like this.

Berkeley is an especially interesting case because it has limited public displays – the museum exists primarily to support the students on campus. And Ray and I knew that a whole bunch of secrets were hidden behind those walls.

The history of paleontology at Berkeley is rich, and most West Coast paleontologists have spent time there, either as students, faculty, or visiting scientists. The state of California had an explosive beginning with the discovery of gold at Sutter's Mill in January of 1848 and the acquisition of California from Mexico in the Treaty of Guadalupe Hidalgo just over a week later. Within a year, the gold rush was on and prospectors, miners, and geologists were streaming into the port of San Francisco. By 1850, it was a state, and by 1860, the state had its own geological survey. By 1874, the fossils discovered by the survey were transferred to the young University of California in Berkeley. That same year, the geologist and paleontologist Joseph LeConte, a refugee from post–Civil War South Carolina, became the first professor of paleontology. LeConte was a devotee of Darwin and an ardent spokesman for the concept of evolution, and he became a close friend of John Muir. His 1878 textbook, *The Elements of Geology*, amply illustrated the fossil record, and he followed that with a popular text aimed at a high school audience in 1888. When I cleaned out my grandparents' home in Fresno in 1978, I found a copy of that very text. It was clear that LeConte was not opposed to religion, but it was equally clear that he did not see any conflict between evolution and religion.

LeConte hired his student John C. Merriam in 1894, and Berkeley was well on its way to dominance of West Coast paleontology for the next century. Merriam did much of the early research on the La Brea fossils and also ventured north into Oregon to the John Day Fossil Beds (famous for its Eocene, Oligocene, and Miocene fossils). Like LeConte, Merriam was a popularizer of ancient life, and in 1915 he published his *History of Life on the West Coast*.

Paleontology at Berkeley was both aided and impaired by a sugar heiress from Hawaii who had a passion for paleontology, and whose fortune would sustain the growth of the program for nearly half a century. Annie Montague Alexander showed up in Berkeley around 1910 and fell in love with fossils. She would join the paleontologists in the field and enjoyed digging. She was also very wealthy and used her funds to influence the workings of the university, often shaping the hiring and firing of faculty and the directors of departments. During that time, the science of geology and paleontology at Berkeley was fractious on the campus, but it was also vibrant, and the string of names of people who worked and studied there forms a big part of the "Who's Who" of West Coast paleontology.

In 1926, during one of those chaotic periods, paleobotanist Ralph Chaney arrived on campus and began to research West Coast fossil plants and the origin of modern West Coast vegetation. California has unique and wonderful vegetation – from the bizarre Joshua trees of the Mojave Desert to the giant sequoias of Yosemite to the oak savannahs of Napa to the towering coastal redwoods of the north coast. Chaney wanted to know how that world had formed. He and his students worked their way across the state and up into Oregon, finding layers of fossil leaves, nuts and seeds, and petrified forests, and they began to unravel the history. Chaney learned that the coast redwoods (*Sequoia sempervirens*) and the giant sequoias (*Sequoiadendron giganteum*) had a rich fossil record. He also found and described many of their extinct fossil relatives.

One of Ralph Chaney's students, Erling Dorf, became a professor at Princeton, where he taught a student named Leo Hickey, who later became my professor at Yale. While my real great-grandfather was from Sweden, my academic great-grandfather was from Berkeley. In the obscure ways of science and paleontology, things that happened a century ago on college campuses still matter to this day.

## Dawn Redwood

By the beginning of the Second World War, Ralph Chaney was considered one of the nation's premier paleobotanists. In 1941, botanists in central China discovered groves of a tree that looked like a coast redwood but wasn't. In fact, it turned out to be a living example of one of Ralph's fossils, and after the war he journeyed to China to see the trees for himself. Indeed, the living trees were identical to his fossil, and in 1948, he named a new species of conifer, *Metasequoia glyptostroboides*. The publication of his find came just a decade after the discovery of the presumed-extinct coelacanth fish living off the coast of Madagascar, so the world was primed to hear about fossil species being discovered alive in remote parts of the world. Chaney became about as famous as any paleobotanist could be, and he made several trips back to China, returning with seeds of his new "dawn redwood." Many of these were planted on college campuses and arboreta around the country, and now, in their eighth decade, many of these trees are towering examples of the relationship between living and fossil plants.

Charles Camp, a vertebrate paleontologist interested in Triassic reptiles, arrived on the Berkeley campus in 1922 and became a favorite of Annie Alexander, whose actions got him appointed head of the museum, while Chaney remained the head of the department. Camp went on to excavate a fabulous Triassic ichthyosaur bone bed near Berlin, Nevada. In 1952, he wrote a book called *Earth Song: A Prologue to History*, in which he used the exquisite pencil drawings of William Gordon Huff – the same WPA artist we met in Chapter 3 – to illustrate the ancient world of the West Coast. Today, this rich history of paleontology continues, and a quick survey of the faculty and staff of the Berkeley museum turned up more than twenty paleontologists. And, importantly, Berkeley is also the home of the National Center for Science Education, a nonprofit group that fights to protect scientists from the attacks of creationists.

I have been corresponding with Berkeley paleobotanists since I was a teenager, and I knew the treasures we would find in Berkeley's well-curated cabinets. By this time, Ray and I had honed our approach for interviewing paleontologists and extracting their tales. It was interesting for me as an active paleontologist to switch into the role of journalist and to interview my colleagues. There was one story that I especially wanted to hear, so we made an appointment to meet with Bill Clemens, the curator of fossil mammals.

Bill is a warm, avuncular man with white hair and beard and a soothing voice. He made his career studying the tiny mammals that lived with the last dinosaurs and those that survived the extinction at the end of the Cretaceous Period. He has mainly worked in the Late Cretaceous Hell Creek Formation of eastern Montana, driving there every summer with teams of students. The fossils he studies are the teeth from animals that were the size of mice. Many of his fossils are as small as sand grains.

Bill was minding his own business in the late 1970s when a geologist, a physicist, and two

chemists from across campus proposed the idea that an asteroid had caused the extinction of the dinosaurs. At the time, Bill probably knew more about the actual fossil record of land animals at the end of the Cretaceous than anyone in the world, and he was surprised to hear their theory. Based on what he knew, he opposed it, suggesting instead that the fossil record showed a gradual rather than catastrophic termination to the Cretaceous. When the team led by Walter Alvarez published their hypothesis in 1980, Bill Clemens rebutted it in print and it was game on.

I had a ringside seat for this fight, as my Yale professor, Leo Hickey, was Bill's friend and colleague (he was also Walter Alvarez's Princeton classmate). The conversation got personal when Nobel laureate and physicist Luis Alvarez, Walter's father, insinuated that paleontologists were not real scientists but were instead little more than "stamp collectors." The next decade was both exciting and painful, as the scientific and personal conflict waged on. Eventually, the existence of the asteroid impact became impossible to ignore and a general sense that it was the cause of the Cretaceous extinctions came to dominate the scientific community.

By the time Ray and I walked into Bill's office, thirty years had elapsed and I was really curious to hear how Bill felt about the whole thing now. He welcomed us warmly and graciously answered our questions about his childhood in California. He had gone to high school with William Huff's daughter and arrived as a student at Berkeley in 1955, just in time to take Ralph Chaney's last paleobotany class. His love of fossil mammals formed when he visited his grandfather's ranch near Torrington in southeastern Wyoming and found fossil bones sticking out of the ground. As a student, he learned that it was possible to use a window screen to sort tiny fossil mammal teeth from loose sand, and soon he was on his way to studying the mammals

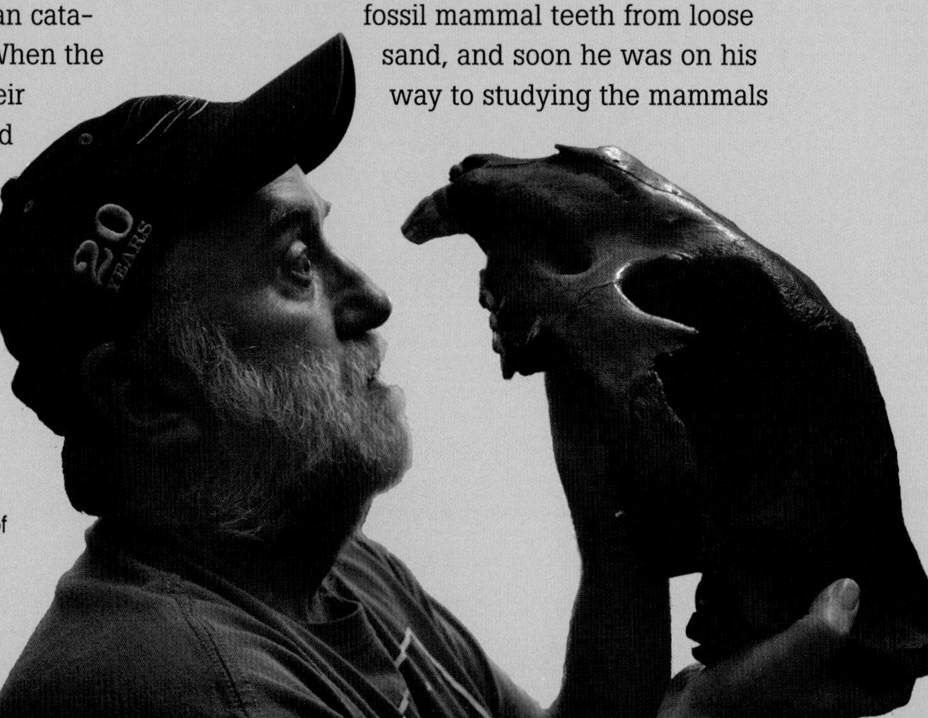


Above: Bill Clemens ponders the tooth of a Cretaceous mammal.

Right: Ray considering his fate as the prey of *Panthera atrox*, the American lion.

## Lupe the Mammoth

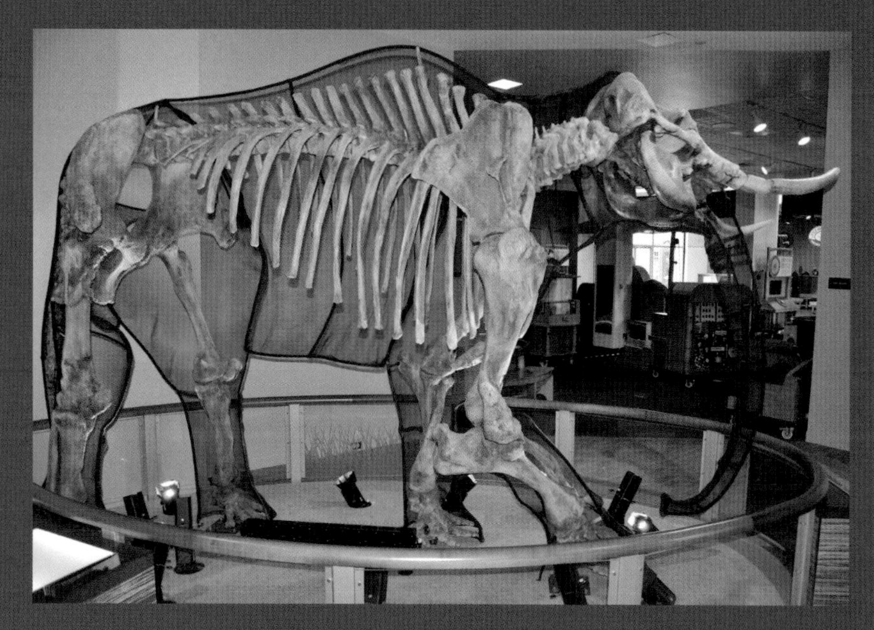

In the summer of 2005, a truck mechanic named Roger Costillo was walking his dog, Jenna, through the middle of downtown San Jose when he – or the dog – spotted a bone sticking out of the bank of the Guadalupe River. It turned out to be a mammoth tusk. The Berkeley crew was called in to investigate, and they excavated a partial skull, a femur, part of the pelvis, pieces of ribs, and several toe bones. The San Jose Children's Museum suddenly had a mascot mammoth named Lupe. Naturally, we stopped into the museum and visited the skeleton. Because it was a partial skeleton, the bones were displayed in glass cases, and the museum had contracted for a reconstructed skeleton to be mounted within a wire mesh screen to show the shape of the whole animal. One thing that jumped out at me was its incredibly long tail. Outside the museum, there was a complete fleshed-out mammoth, which, apparently, had arrived by helicopter. A dog finds a bone and a helicopter delivers a mammoth. Such is paleontology.

that lived with the last dinosaurs. We discussed his long and productive career and wondered if he had finally warmed to the asteroid extinction theory. His responses suggested that thirty years had not been long enough to convince him and that he was still searching for an alternate explanation for the extinction that was more nuanced than a giant rock from outer space.

Pivoting away from this tender topic, I directed our line of questioning to the expeditions that Bill had led to the North Slope of Alaska in search of Late Cretaceous dinosaurs. Ray and I hoped to visit these sites ourselves, and we were full of questions about how we might pull it off. Ray is petrified of bears, and Bill didn't do much to quell his fear. At the time, our Alaska expedition didn't seem very feasible but, as you will see in Chapter 13, it happened.

We spent a few days in Berkeley, harassing scientists and pulling drawers in the endless collection. The more we looked, the more we realized that the Bay Area was loaded with obscure and little-known fossil sites.

The guys at Berkeley told us a quaint story about a gravel pit near the town of Hayward. In 1940, a few fossil bones were discovered in the pit, and a local high school teacher named Wesley Gordon organized a group of seven-to-thirteen-year-old boys to help him dig out the fossils. The boys got good at collecting the bones and they formed quite a team. Soon they were known as the "Hayward Boy Paleontologists," and they found enough cool stuff that they were featured in *Life Magazine* in 1945. They continued working as a group into the 1950s and eventually collected more than 20,000 fossils that included plants, clams, and more than 50 species of animal, including fish, frogs, turtles, rodents, dire wolves, camels, horses, birds, seals, four-horned antelopes (*Tetrameryx irvingtonensis*), deer, mammoths, mastodons, ground sloths, short-faced bears, and saber-toothed cats. In fact, Berkeley paleontologist Don Savage eventually named the Irvingtonian Land Mammal Age (the same age as the fossils from the Fairmead Landfill near Fresno) for the fossils found at this site. Eventually the site was buried under road construction, and today it lies beneath Interstate 680 near Route 238. If you visit the town of Fremont, you'll find Sabercat Historical Park, created to celebrate this discovery.

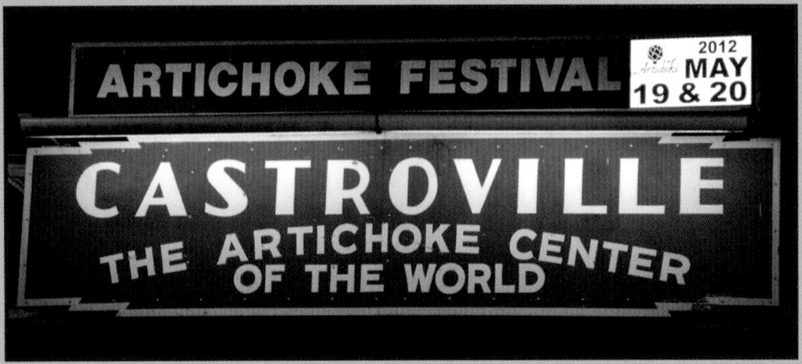

Mammoths and mastodons were big animals, and when they died, they often left behind big bones – bones that are hard to miss and even harder to ignore. They pop out of construction sites and stick out of riverbanks and sea cliffs. If you doubt this, just Google "mammoth discovery" and many recent examples will pop up. Because of their size and obviousness, mammoth and mastodon fossils are relatively common discoveries, and if paleontologists are alerted, they can usually find additional bones from smaller and less obvious animals along with them.

We were now headed to Monterey to see a mural that Ray had designed for a National Oceanic and Atmospheric Administration (NOAA) building at Pacific Grove, and our route took us through Castroville. I have always known Castroville as the Artichoke Capital of the World. I'm a big fan of artichokes, and nostalgia will always bend my coastal drives through Castroville just to see the giant artichoke on Main Street. Had we been there in 1948, we could have seen Marilyn Monroe being crowned as the town's first honorary Artichoke Queen.

This time, however, we had a different reason for visiting. An artichoke grower named Ryan Jefferson had been working his field when he came across a curious fossil. This next day, his cousin Martin Jefferson identified the fossil as a mammoth tooth. They called nearby Foothill Community College and made contact with an archaeologist named Tim King. The college responded and the web lit up with news about a mammoth in an artichoke field.

I had called ahead, and Tim was waiting for us when we pulled up at the agreed-upon mile marker. He had the excitement of a kid and was eager to show us the site. A team had been digging into the slope and were finding more than mammoth. They had parts of a Jefferson's ground sloth (*Megalonyx jeffersoni*) as well as camel, horse, and mastodon. Tim had his archaeologist glasses on and was eagerly searching for evidence of humans or human activity with these Ice Age bones. I was looking through my geologist sunglasses, and we had a healthy debate about the nature of the site.

The crowd that greeted us included archaeologists, farmers, and La Brea Tar Pit paleontology volunteers from Los Angeles. Martin Jefferson was operating the backhoe and he was seeing more than fossils. He was seeing dollars. He had already created a "Mammoth Artichoke" logo and was printing it on his artichoke crates. He and Ryan were happy to see us and gave us Mammoth Artichoke T-shirts.

It seemed that the dueling bands of Californian paleontologists from San Francisco and Los Angeles might be warring over who got to collect the fossils found in the land between the two cities. One of the guys from Los Angeles was an artist named Josh Ballze who was dressed in black leather with all sorts of fossil tools and cameras strapped to his arms and legs. Jurassic Park, meet Mad Max..

Frankly, by this time we were getting a little weary of mammoths, mastodons, and gomphotheres. It seemed like California was infested with various kinds of fossil elephants. It's not the first thing that you think of when you think about the Golden State, but it certainly is a fact when you are on, or in, the ground.

Top: Tim King with an artichoke plant and fossil bone in the foreground and a mammoth excavation in the background.

Middle: Martin Jefferson and Dan Cearley with a box of Mammoth artichokes.

Bottom: Ryan Jefferson and a Mammoth Artichoke T-shirt.

We had some genuine West Texas barbecue in Castroville and drove on to Pacific Grove where Ray's drawings were being displayed in the Pacific Grove Museum of Natural History. This museum opened in 1883, not long after California became a state. Even the Smithsonian, which had opened to the public in 1855, did not open the US National Museum until 1881, and many of the other great museums of the United States were founded in the thirty-year window between 1880 and 1910. It was the golden age of natural history and the peak of hobbyist interest in the natural world. In this time before the invention of cars and planes, everyone had the experience of traveling by horse and buggy (or wagon or stagecoach) and, as a result, knew the anatomy of a horse from firsthand observation. Now we talk about horsepower in reference to engines, forgetting that real horse power is how we used to get around. Back then, there were clubs that collected shells, butterflies, bird eggs, beetles, wildflowers, rocks, fossils, and all sorts of other natural curiosities. There was a deep and pervading belief that nature was worth serious contemplation.

The Pacific Grove Museum of Natural History is a sweet little museum that takes you back to a simpler time when beauty and nature were more of a day-to-day avocation for people. Pacific Grove itself sits at the southern edge of Monterey Bay, and its quaint neighborhoods abut rocky headlands in a way that puts residents right next to tide pools. Around the corner, and just to the south, are the seaside village of Carmel, the rocky Point Lobos, and the endless cliffs of Big Sur.

## The Legacy of John Muir

**John Muir, the Scottish naturalist, had made his way to California in the late 1860s and was living in Yosemite at the end of that decade. It was his writing and his persuasive promoting of the beautiful valley that eventually drove the US government to designate Yosemite as a state-managed park in 1890. Two years later, Muir worked with Berkeley paleontologist Joseph LeConte and Stanford ichthyologist and paleontologist David Starr Jordan to start the Sierra Club, one of the first environmental protection organizations in the world. Teddy Roosevelt, himself a naturalist and a collector, visited Muir in Yosemite in 1903 and because of this visit, named Yosemite as a National Park in 1906.**

This is holy ground for me. My parents honeymooned in Carmel, and we returned there many times throughout my life. We would walk the sandy beaches and play in the driftwood. The tide pools at Point Lobos are like candy for kids; each one has its own little aquarium world of anemones, crabs, and sculpin. When I was twelve and we were visiting Carmel, I found myself in an amazing rock shop staring at something I had never seen and could not imagine – a perfect foot-long fossil fish on a creamy slab of rock. I know now that the fossil was collected in a quarry in the Fossil Basin of southwestern Wyoming and that its name is *Mioplosus*. At the time, I could not believe that it was real. I really wanted that fossil badly, but it was quite expensive. I remember a price of $200, an incomprehensible sum. After a lengthy, and I imagine teary, negotiation, my mom actually purchased it for me. Forty-five years later, I still have that fossil, and it makes me think of tide pools, the California coast, and my mother.

For a kid – or anyone – who loves finding things, rough shorelines are an endless treasure trove. It turns out that the shore by the Pacific Grove Museum is no exception. To the south of the museum, along Big Sur, there is even a place called Jade Cove, where pieces of gem-quality jade occasionally wash ashore. While Ray and I were out on the museum's patio enjoying the reception for his show, I noticed a rock that took my breath away. It was a seal-sized chunk of jade sitting on a pedestal. In that glance, my love of discovery, rocky shorelines, and museums was fused into a single glossy-green, gorgeous rock.

Ray's show was great. He had drawn a series of images that told the story of Monterey Bay and its everlasting relationship with blue seas and green seas. The blue water is the warm equatorial water that brings tropical fish like tuna, marlin, mackerel, swordfish, moonfish, sharks, anchovies, and sunfish. The green water is the cooler water from the north that brings salmon, sardines, ling cod, rockfish, cabezon, auklets, and sea lions.

Monterey Bay is where the blue water meets the green water, and as a result it is one of the most interesting spots on the entire West Coast. The anchovies bring the whales – humpbacks and even blue whales. Meanwhile, the California gray whales cruise past on their annual migration from

Baja California to the Bering Sea. Add to that the fact that the deep-sea Monterey Canyon nearly makes it to the shoreline near Moss Landing, and you have a place where the truly deep Pacific Ocean is remarkably close to the beach where people surf. North, south, green, blue, shallow, and very deep, Monterey is a place where the fish once supported a thriving commercial fishery. Today, however, the Monterey waterfront is dominated by tourists who come to see the Monterey Bay Aquarium – one of the best in the world.

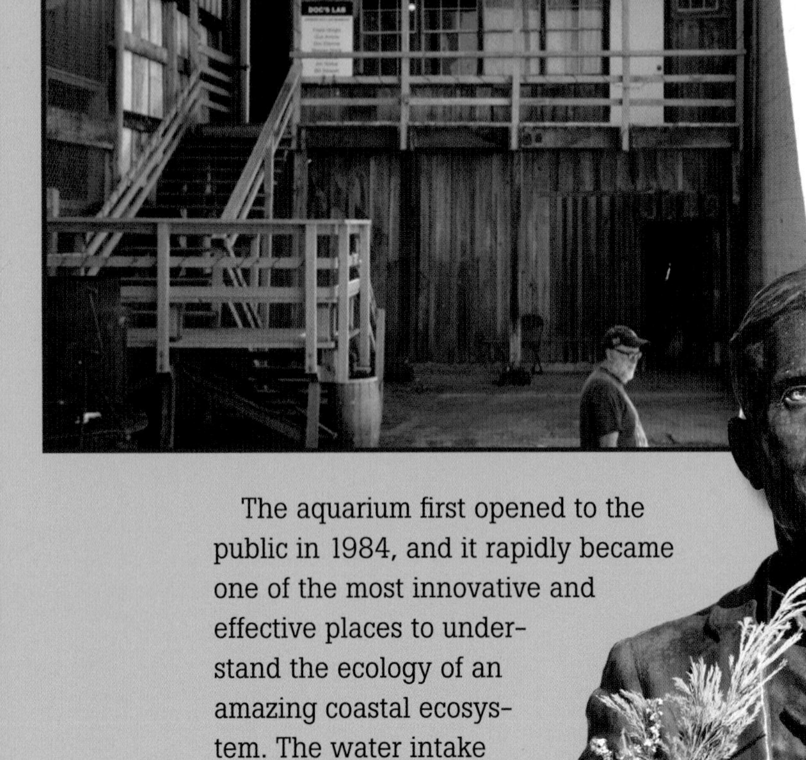

The aquarium first opened to the public in 1984, and it rapidly became one of the most innovative and effective places to understand the ecology of an amazing coastal ecosystem. The water intake pipes from the aquarium reach far out into the bay, and the water that comes into the many tanks is cold, clean, and completely saturated with marine life. Many of the tanks are populated by microscopic marine larvae that treat the aquarium as an extension of their ocean home. Everybody who works at the aquarium seems to be a trained marine biologist, and you can barely buy a T-shirt without learning about the biology of a squid.

Ray is a West Coast fish T-shirt guru, and he has had a long relationship with the scientists and aquarists at the Monterey Bay Aquarium. His ability to render the denizens of the Pacific depths is matched by their ability to display the real thing to a massive and adoring audience. After leaving Pacific Grove, we wandered around the aquarium enjoying the sharks, herring, and jellyfish and watching the crowds experience the life of the Pacific.

The reason the Monterey Bay Aquarium is such a recent addition is that it is located on a street known as Cannery Row, right on the spot where Knut Hovden opened a sardine cannery in 1916. Monterey was a major fishing port. It was here, in 1970, that I lowered a chopped squid from a dock and caught my first fish. It was also here, in 1973, that the sardine fishery crashed because of overharvesting and ended more than a century of fishing.

The story of this particular fishing port was burned into the American psyche by author John Steinbeck

Left: Ray wanders past the tanks behind Ed Ricketts' Lab in Monterey.

Right: The bronze Ed Ricketts is rarely without fresh flowers.

who lived in Pacific Grove in the 1930s and '40s, and who observed and fictionalized the bustling sardine fishery and the people that populated it. In his 1945 novel *Cannery Row,* he told the tale of a curious and charismatic marine biologist named Doc who lived on Cannery Row and who was the soul of the community. Doc, it turns out, was a real person and Steinbeck's best friend, Ed Ricketts. Without formal scientific training, Ed was a different kind of scientist. He made his living by collecting coastal creatures from tide pools, mud flats, and rocky shorelines and selling them to schools and laboratories. His workshop, the Pacific Biological Laboratories, still stands today, at 800 Cannery Row, in the shadow of the Monterey Bay Aquarium. Ricketts was one of the first scientists to think in terms of marine ecosystems rather than basic taxonomy, and he was well ahead of his time. In 1939, he and Jack Calvin published *Between Pacific Tides: An Account of the Habits and Habitats of Some Five Hundred of the Common Conspicuous Seashore Invertebrates of the Pacific Coast Between Sitka, Alaska, and Northern Mexico.* Still in print, and now in its fifth edition, the book is the bible for tide poolers and a testament to the fact that some of the best science comes from the edge of science.

Back in 1940, Ed Ricketts and Steinbeck rented an old sardine boat named the *Western Flyer* and took a six-week trip from Monterey down the California coast to the tip of the Baja Peninsula and up into the Sea of Cortez. Ostensibly, the goal was to collect specimens for Ricketts's business, but the resulting book, *Sea of Cortez: A Leisurely Journal of Travel and Research,* coauthored by Steinbeck and Ricketts in 1941, told a much richer tale of friendship and philosophy. In 1932, Ricketts traveled to Sitka, Alaska, and in 1946 he explored the west coast of Vancouver Island and the Queen Charlotte Islands (now Haida Gwaii). In 1948, Ricketts was killed when the car he was driving was struck by a train just a few blocks from his lab on Cannery Row. In 1951, Steinbeck published *The Log from the Sea of Cortez*, an updated story of the 1940 trip without Ricketts as an author and without his long lists of collected specimens. It did include a lengthy preface that was a loving dedication to, and memorial of, Ed Ricketts.

Bobby Boessenecker holding a sea cow bone amidst suburban trash at the site of Repenning's rookery.

*The Log from the Sea of Cortez* was very much on our minds when Ray and I were exploring the American West in preparation for our 2007 book, *Cruisin' the Fossil Freeway: An Epoch Tale of a Scientist and an Artist on the Ultimate 5,000-Mile Paleo Road Trip.* The idea of an artist and a scientist taking a long fossil-focused road trip was something that seemed right in line with the spirit of the *The Log from the Sea of Cortez.* And it was Ricketts's unachieved ambition to visit the entire coast from Baja to Alaska that tossed down

Fossil-studded boulders on the beach at Capitola.

a marker that defined the 4,000-mile span of our West Coast fossil journey.

It was with this background that we approached our visit to Monterey. I had contacted Jim Hekkers, the vice president of the Monterey Bay Aquarium and an old friend from Denver. Jim had one thing that we wanted very badly: the key to the Pacific Biological Laboratories. After Ricketts's death, his lab had changed hands a few times, but its owners had never really scrubbed the place of his presence. It is now owned by the city of Monterey but is not open to the public. It just sits there on Cannery Row, a modest reminder of an amazing man and a different time.

Being in the lab was a moving and memorable experience. It was not just a lab, it was Ricketts's home, and the trappings of his trade and his habits were still quite visible despite the sixty-two years that had passed since his death. Down the street, next to the railroad crossing that killed him, there is a small bronze bust of Ricketts. Look closely and you can see his bronze hand lens, the sure sign of a biologist or botanist. It is said that there are always fresh flowers in his hands, and it was true for the week we spent in Monterey. We drove out of town with a strong sense that our exploration of the West Coast was well worth pursuing.

I had been searching the internet for interesting individuals who were an active part of the West Coast fossil scene, and one day I came across a blog called *The Coastal Paleontologist*. Blogs are hit-or-miss operations, but this one was a jackpot. Written by a guy I had never heard of, it was full of facts and useful information, and it was clear that this guy knew what he was talking about. I emailed him and asked if he would mind if an artist and a scientist dropped by for a visit. He agreed, so Ray and I drove to Foster City south of San Francisco and arrived at the home of Bobby Boessenecker. Actually, it was his parents' home.

We pulled up to a typical suburban California home and immediately noticed concretions in the garden. Sticking out of the concretions were fossil bones. We were at the right place. Bobby answered the door and welcomed us into a typical living room with atypical contents. Bobby's dad is a historian and his mom is a judge. Bobby had just spent the previous weekend in a kayak along the coast north of Point Reyes surveying a Pliocene rock layer known as the Purisima Formation. The Pliocene is only 3 to 5 million years old, so you would expect to find fossils of ani-

mals that are pretty similar to those alive today. Much to my delight and amazement, Bobby had found a block of rock with the back side of an imagotarine walrus skull sticking out of it. He told us about a thirteen-year-old kid named Forrest Sheperd who had found a near-perfect skull of this animal in Santa Cruz. Then he showed us a host of treasures that included dolphin skulls and fossil walrus tusks the size and shape of *Smilodon* sabers. Bobby was just finishing his master's degree on the Purisima Formation, and he agreed to take us for a drive to see some outcrops and perhaps a fossil or two.

We took off and headed over to "the 280" (Californians refer to their highways by their numbers) on our way over the Santa Cruz Mountains. I had worked for the US Geological Survey in Menlo Park back in the early 1980s, and this was familiar ground. As we drove past Stanford, Bobby pointed out a gully that had yielded fossil whales and another that was full of fossil barnacles, and he reminded us that Stanford's first president, David Starr Jordan (co-founder of the Sierra Club), had been an ichthyologist and a

Upper left: Fossil whale bones.

Right: Bobby Boessenecker inspecting a fossil whale skull in a beach boulder.

Left: The surfers oblivious to the fossils at Capitola.

Opposite: Skeleton of the California gray whale (*Eschrichtius robustus*).

fossil shark guy who had worked at Sharktooth Hill back in the 1920s. Then we crossed over the Stanford Linear Accelerator (SLAC), the longest, straightest structure in the world. Built in 1962, this is a 2-mile-long ongoing physics experiment and particle collider. I remembered a story about a skeleton that was discovered during the construction of the facility and called to inquire about visiting hours. They told me that they allowed the public in once a month, but this was not that day. Bobby had also heard about the skeleton but had never been able to see it himself.

Halfway over the Santa Cruz Mountains, we pulled off at Scott's Valley and visited forested slopes that contained fossil sand dollars and shark teeth. Driving down a winding forested road, we pulled over to the side and looked at a spot where US Geological Survey paleontologist Charles "Chuck" Repenning had collected parts of several Miocene walruses (*Imagotaria downsi*). Bobby called it Repenning's rookery. In a place called Laveage Park, he showed us a spot where Howard University's fossil sea cow expert Daryl Domning had collected the two fossil sea cows *Dusisiren jordani*, a mother and a calf that are now on display at the Los Angeles County Museum.

Many of these places were spots where people had found and removed fossils, and there was nothing there for us to see but the sandy outcrop. But in one spot, Bobby showed us a sea cow skeleton still in the bank. It is amazing to realize that these poison oak- and pine-covered hills are actually uplifted Pliocene seafloor.

The next stop was Capitola, a quaint little seaside village just south of Santa Cruz. The main street dead-ended at the ocean, and the town was all bars, surfer shops, and taco shacks. The beach was made of big, flat boulders, and everybody was surfing. It was sunny and breezy and a perfect California day. But it was even more perfect than that. The beach was made of fossils. The boulders were studded with fossil clams so obvious that you would have to be blind to miss them. Bobby pointed out more subtle features like fossil crab claws, and once ours eyes adjusted to the color of the rock, we realized that the boulders also contained huge fossil bones. The waves had rounded the rocks, and because the bone itself was as hard as the rock, it was rounded right along with the rock. We had to look thoughtfully at the smooth surface of the boulders to recognize the embedded bones; these were whale fossils from big whales – baleen whales. The beach was made of fossil whales. And all around us, people were surfing.

Next, we drove through downtown Santa Cruz, past spots that had yielded fossil walrus and sea cows. We stopped in the Santa Cruz Museum, and sure enough, the place was full of local fossils. Around the corner at the Long Marine Lab there were two mounted whale skeletons, a blue and a gray. Looking at those massive skeletons, it was hard to imagine the work it would take to chop a full fossil whale out of rock. It is said that vertebrate paleontology has two types of fools, those that work on the giant long-necked sauropod dinosaurs and those that work on fossil whales. I actually think the whales might be worse than the dinosaurs.

We bypassed the elephant seal rookeries at Año Nuevo State Park and stopped for dinner and artichoke soup at Duarte's Tavern in Pescadero. Then we dropped Bobby back in Foster City and headed to Marin County. The next day, I dropped Ray at the San Francisco airport, and he headed back to Alaska. Then my dad picked me up and we headed north to Seattle.

Dad and I drove by way of Calistoga, California, where there is a truly remarkable petrified forest. I have vague memories of visiting this forest when I was a kid, but to see it as a trained paleontologist was something entirely different. Here is a forest that had been blown down and buried by a volcanic eruption about 3.4 million years ago. Entire trees are preserved, but what is truly amazing to me is that this is one of the very few places I have ever seen fossilized bark. Curiously, the fossil forest lies partially buried in a modern forest; huge fossil sequoia and Douglas fir trees lie petrified just below the floor of a living Douglas fir forest. It is one of those very rare places where a fossil site preserves an ancient landscape that is not very different from the landscape that exists today.

It turns out that my real family also has a history with the forest. My dad's aunt Ethel had lived on a ridgetop in nearby St. Helena, California. She was a devout Seventh Day Adventist who had lived for a while in Hawaii and had a greenhouse full of the most exquisite orchids. By coincidence or design, St. Helena was also the home of Ellen G. White, one of the people who founded the Seventh Day Adventist religion in 1860. A prolific writer, White became the prophetess of the church and the arbiter of its moral framework. She moved to St. Helena in 1900 and died in 1915, so she would not have been there when the petrified forest was discovered, but she would have been aware of it. This was the time when the Seventh Day Adventists were becoming vividly aware of the significance of the fossil record and were struggling to reconcile it with their seven-day version of the Earth's creation.

Despite the fact that I visited my great-aunt regularly and that she was aware of my burgeoning interest in fossils, she never mentioned the site to me, nor did she take me there. I didn't think much of this until recently when my second cousin, John Drage, was visiting and recalled his visit to Ethel in the 1970s. He said he had been startled and scared when Ethel, normally a very cautious driver, closed her eyes and hit the accelerator while driving past the Petrified Forest. He realized that she would literally not let herself see evidence of a fossil world that contradicted the world of her Bible. She would not see what she did not believe.

The same cousin related a story of my grandfather Elmer whom he visited in Fresno after hiking the Grand Canyon. John was talking about his hike and going on in some detail. Elmer fell asleep, and John continued talking to my grandmother. At the point in the story where John made it to the bottom of the canyon and saw outcrops of the 1.73-billion-year-old Vishnu Schist, Elmer jolted out of his slumber and roared, "Six Thousand Years!"

My dad had attended the SDA college in nearby Angwin, but he had never visited the fossil forest before. We had a delightful stroll in the fossil forest in the living forest and chatted about the curious coincidence of family history, scientific history, geology, botany and religious dogma.

### Petrified Charlie's Forest

**The Calistoga Petrified Forest was discovered in 1870 by a Swedish homesteader who came to be known as "Petrified Charlie." It was visited that year by O. C. Marsh, the famous Yale paleontologist, who was visiting the Bay Area, and the next year published a paper on the petrified forest. The noted author Robert Louis Stevenson also visited the site and wrote about it in his book *The Silverado Squatters.* Paleobotanist Ralph Chaney described the site in his research on the origin of redwoods and ended up assigning it to Erling Dorf (my academic grandfather), who wrote part of his PhD about it.**

Far left: My dad drove his MG TF through a redwood tree in 1955 and we drove through it again in 2010.

Upper right: Marine shale with the spot where I found my ammonite.

Lower right: The ammonite in hand.

As we continued through the northern end of California's Central Valley, I was reminded of the many outcrops of Cretaceous marine shale in the area – outcrops that hold one of the most interesting ammonite faunas in North America. Exposures of 80-million-year-old marine shale are found in central Baja, San Diego, northern California, southern Vancouver Island, and central Alaska. In each one of these places, extraordinarily beautiful ammonites are preserved in concretions that weather out of the shale. The string of sites speaks to the nature of the 80-million-year coastline and the challenges of reconstructing it. In northern California, the presence of booby-trapped marijuana plantations makes it both scary and dangerous to hunt ammonites.

All I could think of was ammonites. We stopped at a roadcut and found that it was made of marine shale. I knew that there would be an ammonite if I looked hard enough, and after thirty minutes of close inspection I finally noticed the edge of sweet 4-inch ammonite poking out of the hill. I didn't have tools with me so I used a sharp stick to pry the fossil out of the rock. With the ammonite mission accomplished, we continued our trek north.

A year later, I was back in Menlo Park. This time I had taken pains to contact the officials at the Stanford Linear Accelerator to make a better argument about why I should be allowed to visit their well-guarded visitor's center. I was traveling with Jane Lubchenco, a marine biologist and the head of NOAA who was temporarily based at Stanford. This time, permission was granted, so

Jane and I drove to SLAC. We were met by officials who welcomed us into a small building that had exhibits about the construction and scientific successes of the facility. And then, I saw what I had been waiting to see for so many years, the magnificent mounted skeleton of a *Neoparadoxia repenningi*, a giant extinct desmostylian. Until Larry Barnes had exhibited his Orange County specimen in the new mammal hall that opened at the Los Angeles County Museum in 2010, there was no other skeleton like this on exhibit on the West Coast. This one had been on display since 1998 but only behind a very serious restricted entry gate.

In 1961, Stanford University made a big play in particle physics by approving the construction of the world's largest linear accelerator. They appointed the distinguished physicist Wolfgang Panofsky, or "Pief" as he was called by his students, as its first director; construction would take five years. In many ways, this was an odd place to build a very long, extremely straight, and perfectly flat building. You might think that the plains of Nebraska would be a better choice, but Stanford is not in Nebraska, it is near the San Andreas Fault and at the edge of the Los Altos Hills. For this reason, the first step of the construction was to dig a very long, very straight trench into the local bedrock. And when you dig into fossil-bearing bedrock, you find fossils. The workers had been encountering shark teeth, whale bones, and parts of porpoises and pinnipeds, but on October 2, 1964, a bulldozer sliced into something amazing: a complete skeleton of something that nobody could recognize. Chuck Repenning was called up the hill from his office in Menlo Park to inspect the site and, recognizing that it was likely some sort of desmostylian, he quickly obtained permission to excavate it. He entrained the aid of a few students and volunteers, including Adele Panofsky, the director's wife.

In a 1998 article, Adele described the find: "The specimen had been buried lying on its back on the sand of the sea floor. The ribs had fallen to the sides before burial and at least two dozen isolated shark teeth were found in the skeleton or very near to it." It took the crew six weeks to excavate the skeleton, and by the end of the dig, Adele had found her lifelong obsession.

Over the next twenty-four years, she devoted her free time to cleaning up the fossils, transferring them eventually to the collection at Berkeley, and then devising a way to reconstruct the skeleton and display it at SLAC. She visited museums around the world to interview the few scientists who knew anything about these animals. And she studied the techniques of mounting fossil skeletons. The bones were fragile so she worked with plaster casts rather than trying to mount the actual fossil. Eventually she succeeded, and in 1998, she finished the display, complete with a Miocene marine mural painted by her granddaughter Catherine.

The specimen came to be known as the "Stanford *Paleoparadoxia*," and in 2007, Larry Barnes and Daryl Domning officially named it *Paleoparadoxia repenningi*, or "Repenning's Old Paradox," after Chuck Repenning, who had been tragically murdered in Denver in 2005.

I was overjoyed to finally see this hard-to-access fossil display. Jane and I engaged in an animated discussion about its unlikely home and its scientific significance. As the head of NOAA, Jane's responsibilities included managing the National Marine Mammals Laboratory, and she was thus the highest-ranking marine mammal

Adele Panofsky and her magnum opus, the *Neoparadoxia repenningi* skeleton mounted in the visitor's center at the Stanford Linear Accelerator.

scientist in the country. And here was a fossil marine mammal from the only major group of marine mammals to have ever become extinct. The desmostylids first appeared around 30 million years ago and became extinct about 10 million years ago. For their 20-million-year turn on Planet Earth, they lived only in the North Pacific, with their range stretching from Baja California to Alaska to central Japan.

While we were engaged, an elderly woman in blue jeans wandered into the visitor's center, and our hosts introduced us to Adele Panofsky. Never in 20 million years had I expected her to show up, but the SLAC staff had warned her of our arrival, and here she was. She is truly a desmostylid-obsessed woman, and she was so excited to tell us the story of discovery and resurrection. Her hands had grown quite arthritic, and she was proud to show us that her fingers were like those of her fossil. That visit happened in June of 2013, and three months later, Larry Barnes published another paper that compared Adele's fossil to the Orange County one in Los Angeles and adjusted the name to *Neoparadoxia repenningi*, or "Repenning's New Paradox."

In the time since our visit, Adele's *Neoparadoxia* has been dismantled, and there are plans to mount it in a new museum in San Mateo where it will be more accessible to the public. Slowly but surely, our beloved desmostylians are creeping into the public eye.

OREGON
FOSSIL MAP
Aglyptorhynchus
Extinct Billfish
Astoria
Seaside
Columbia River
Columbia Maritime Museum
Extinct Basking Shark
Kelso
Cougar
Trout Lake
Glenwood
Mabton
Prosser
Richland
Columbia River
Desmatophoca
Cannon Beach
Portunites Fossil Crab in nodule
St. Helens
Turritella
Stonehenge replica
Hermiston
Echo
Fossil Palms
Vancouver
Cape Meares
Behemotops a Desmostylian
Rockaway Beach
Tillamook
Ratfish Egg Capsule in nodule
Portland
Beaverton
Tigard
Tualatin
OREGON MUSEUM of Science + Industry OMSI
Bonneville
Hood River
Mosier
The Dalles
Wasco
Explorer John Day stripped naked here
Promerycochoerus
Mount Hood
Orohippus
four toed horse
Hemipsalodon
Enaliarctos
Erratic Mammoth Sculpture
Tualatin Mastodon at the library!
Dayton
Woodburn
McMinnville
A Derpglo
Shovel Tusker
Condon
Banana Seeds
Cynarctoides "little beardog"
Maupin
petrified wood
Aetiocetus
Sitka Center for art + ecology
Otis
Lincoln City
Aetiocetus
Scott Mills
Salem
Columbian Mammoth and Mammoth tracks
Pristichampsus
Clarno Nut Beds
Menagerie Wilderness Flora
Clarno Unit
Fossil
Oregon Paleo Lands Institute
Skull of Pogonodon lassoed
Daeodon
Ukiah
Kolponomos
Cornwallis
Harlan's Ground Sloth
Terratorn woodburnensi
"running croc"
Painted Hills
"Mitchell's Monster"
Spike Toothed Salmon
Gateway
Plesiosaur
Newport
The Oyster Bear
Oregon Coast Aquarium
Giant "Terror" Bird
Corvallis
from Woodburn
Madras
Metolius
Oncorhynchus rastrosus (Smilodonichthys)
Unit
Mitchell
John Day Fossil Beds National Monument Sheep Rock Unit
Thomas Condon Paleontology Center + Visitors Center
"Pugnosed" dolphin
Atuia Nautilus
Tylocephalonyx the dome-headed Chalicothere
Museum of Natural and Cultural History
Condon Fossil Collection
humerus
Sisters
Prineville
Redmond
Cretaceous Pterosaur
Archaeotherium aka "Hell Pig"
Simocetus
Sea Lion Caves
Eugene
Tumalo
Bend
Trilobites
Jurassic Marine Crocodile
Florence
Hexanchus Shark teeth
Goshen
Odontaspis Shark
Reedsport
Cottage Grove
Curtin
Oakridge
La Pine
Ammonites
Lava Cast Forest
Conodonts
"Bernie" the Thalattosaur
Pontolis magnus Giant "Killer" walrus
Tropical Goshen Flora
Brothers
Hampton
Riley
Burns
Hines
Lakeside
Ammonites
Metasequoia Official State Fossil of Oregon
Idleyld Park
Wally the Whale
"Hole in the Ground"
Pleistocene Horse and Flamingo
North Bend
Fossil Green Sea Turtle
Coos Bay
Sitkum
Diamond Lake
Fossil Lake
Pliocene Peccary
Beaver Marsh
14,000 Year Old Human Coprolites
Summer Lake
Chinook Salmon fossils
Stinking Water Petrified Wood
Bandon
Langlois
baleen whale
a very big pleistocene
Hoploparia "lobster"
Crater Lake
Crater Lake National Park
Union Creek
Playa
Paisley 'Poo'
Giant Camel
fossil snail
Cape Blanco
Port Orford
Shady Cove
Paisley caves
Paisley
Lake Abert
Dire Wolf
Slipper Shell
Crepidula
Prehistoric Gardens
World's Largest Caveman
Rogue River
Bison antiquus
Paramylodon ground sloth
Valley Falls
Trout Creek flora
Grants Pass
Gold Beach
White City
Upper Klamath Lake
Columbian Mammoth
Beatty
Bly
Short Faced Bear Tracks
Ichthyosaur
Medford
Talent
Ashland
Aspen Lake
Klamath Falls
Lakeview
Duckbilled Dinosaur Oregon's only chunk of dino... so far
Cave Junction
Oregon Caves National Monument
Oregon Shakespeare Festival
Keno
Bonanza
Brookings
Worden
Malin
Dog Lake
Goose Lake
Arctodus
California
Nevada

WASHINGTON
Walla Walla
Tempskya Cretaceous Tree Fern
ART-SCIENCE
CBR
CURIOSITY BASED RESEARCH
Snake River
Elgin
Harpoceras ammonite
Airport Lane fossil find
Arctodus short faced bear
Bison antiquus
corals
LaGrande
Columbian Mammoth
Hell's Canyon
short legged rhino
Corals
Tree fern
Teleoceras
Shastasaurus
Haines
Baker City
New Meadows
Nimravids
Agriochoerus
Moropus
Gomphothere
(none the) Weiser
IDAHO
Ictalurus Miocene Catfish
Ontario
Vale
GIANT MINNOW JAWS
Adrian
"on the ground like a wild..."
Succor Creek flora
Snake River
Dromomeryx
Jordan Valley
Ptychocheilus
Burns Junction
Amphicyon
Pliohippus
IN RUTILA VERITAS
SEARCHING FOR THE TRUTH AT THE END OF A SHOVEL
Paleo-pelican
McDermitt
Ray Troll 2013
COLORING: Grace Freeman 2013

# OREGON

# 5 TWO MARLIN AND ONE BANANA

Sometime late in October of 1961, the November issue of *National Geographic* landed in our mailbox in Seattle. I don't remember that moment because I was only thirteen months old at the time, but that event was to have a significant impact on my life. Fortunately for me, my parents, and everyone we knew, held on to their *National Geographic* magazines, filing them in stacks and shelves because they were different from other magazines. Sometime around four years later, I began to mine the yellow trove, flipping the pages and gawking at the pictures. I don't remember how old I was when I first dug into the October 1961 issue, but I can sure remember doing it. On pages 718–719, there were three pictures that burned holes in my brain. The first one showed a kid … who looked like me … peering across a rock … watching a guy named Douglas Emlong … who looked like James Dean chipping away at a fossil scallop shell with a rock hammer and a

A pair of 20-million-year-old marlin and a bunch of bananas frame a lone fossil hunter on the stormy Oregon coast.

center punch at a place called Moolack Beach in Oregon. The second image was the toothy snout of a fossil sea lion smiling at me out of a rock. The third thing was something beautiful that I had never seen or imagined before. It was labeled as the tooth of a sea cow, but I now know that it was a *Desmostylus* tooth. I wanted to be that guy. I wanted to find those fossils. And I really wanted that *Desmostylus* tooth.

The article was a travel piece about the Oregon coast, and I remembered being very annoyed that the whole article wasn't about the fossils of the Oregon coast. Fortunately for me, we lived in Seattle and my grandparents lived in California, and my parents liked driving between the two places along the coastal route. So, it wasn't long after I read the article that we were driving through Lincoln City, Oregon, headed south. I had done my homework and knew that we were approaching Moolack Beach. It was a sunny day, and it was easy to convince my mom to give me a thirty-minute fossil break. She pulled over to the side of the road and turned me loose. She and my sister stayed in the car.

I scrambled down a short, steep, muddy trail to the beach. The sandy beach was backed by a series of small crumbling headlands, and I wandered into a field of engine-sized freshly fallen boulders. Fossils were everywhere. Each boulder was speckled with the cross sections of clams, and in many places, the clams themselves stuck out of the rock, just like ripe apricots waiting to be picked. I could not believe how easy it was. I quickly filled a bag with choice specimens, then

A young Paul Zahl watching Doug Emlong chip a fossil scallop shell from the rock in a picture from the November 1961 issue of *National Geographic.*

Image courtesy of *National Geographic*

dumped it out and started again. You hear about the kid in a candy store, but how often do you see the kid behind the counter at a candy store?

It was funny, but the euphoric feeling faded quickly because I was not finding any fossil sea lions or *Desmostylus*. I was no Doug Emlong. I climbed back up the bank and we carried on toward California.

A few years later, when I was a junior member of the Bellevue Rock Club, I met a woman named Jane Cushing who said that she had found some skulls on the Oregon coast. I kept bugging her to show them to me, and she finally agreed to bring them over to my house. Jane was in her late fifties, and my memory of her today makes me think of Bart Simpson's blue-haired aunts. She pulled her 1968 Chrysler into my driveway, and I ran out to meet her. She lit up a cigarette and opened her trunk.

There, lying on a blanket, were six round rocks, each the size of a large, slightly flattened grapefruit. At first, I thought that Jane was crazy and that she had mistaken round rocks for fossil skulls. People do that more often than you might expect. As I looked more closely, I realized that she had something truly amazing. Each rock was, in fact, a skull.

Imagine that a marine mammal dies and sinks to the bottom of the sea. Crabs, amphipods, and microbes quickly consume the skin and flesh, and eventually the skull is buried in seafloor mud. Over time, the mud hardens to bedrock, and something about the chemistry of the rotting skull causes the mud that infills it to become even harder than the surrounding mudstone. Then the bedrock is uplifted into the surf zone where waves attack it and break it into blocks that tumble in the surf and become round beach boulders. The boulders that are cored by skulls are tougher, and they last longer in the smashing surf. Eventually, the skull begins to peek out of the boulder. The snout, eye sockets, and skull crest come into view. Waves exhume the long dead of the Pacific, cast them on the shoreline, and literally buff them into the here and now where anyone can come along and find them if

Kent Gibson doing what he does every morning, scanning the beach for freshly exposed fossils.

they have a well-tuned search image.

That was the last time that I ever saw Jane, and I've often pondered what happened to those skulls and what it was that inclined a chain-smoking, middle-aged woman to take the time to share her fossil finds with an eleven-year-old kid. And now I think I know. Fossils are a window into a magical world that just happens to be real.

### The Fossil Mailman

**Portland has never had a natural history museum, but the Oregon Museum of Science and Industry (OMSI) has supported paleontology in a variety of ways, the most important one being that they have run a paleontology field camp at Camp Hancock in central Oregon since 1951. In fact, I know of a surprising number of professional paleontologists who experienced their first wild-caught fossil at Camp Hancock. The camp is named for Alonzo "Lon" Hancock, a Portland mailman who became obsessed with the fossils of central Oregon in 1931 and led both professionals and kids to his various discoveries. He built an impressive home museum that delighted neighborhood kids and visiting professionals. When he passed away in 1961, he left more than 10,000 specimens to OMSI, and they named the camp for him.**

Forty years later, in February of 2012, I landed at the Portland airport to meet Ray and begin our tour of Oregon. As I walked off the plane, there was a guy with long blond hair, a well-trimmed beard, and a biblical robe who looked exactly like a 1950s Jesus playing an amplified violin in the concourse and flooding out the loudspeakers with insanely loud new age music. A lone woman was sitting cross-legged and meditating in front of him. There was a card table full of CDs for sale. I tweeted, "Just landed in Portlandia" and received the immediate response, "Put a bird on it."

This city at the mouth of the Willamette River has ever been the kid sister to Seattle but now is awash in hipsters and is pushing out its own vibe. Oregon, on the other hand, is a largely rural state with huge expanses of sparely inhabited mountains, deserts, and a fantastically accessible coastline. Highway 101 runs faithfully along the Pacific Coast for its entire 300-mile shoreline.

Ray had rented a car for the curiously low price of $7 a day, and we headed off to our destination, the Sitka Center for Art and Ecology in Otis, Oregon (more about this in a moment). This seemed like the perfect place from which to base our excursions. It was pouring rain and the traffic was awful as we slowly crept through Tigard (home of MacKenzie Smith, the middle schooler who led the successful campaign to designate *Metasequoia* as the Oregon State Fossil) and McMinnville (home of the McMinnville Mammoth site and the Yamhill River Pleistocene Project) before the traffic let up, and we were able to cross the Coast Range on Highway 18 and arrive in Otis, a tiny four-building town just north of Lincoln City, where we had a dinner appointment with Frank and Jane Boyden, the founders of the Sitka Center. Both their home and the Center are located in the valley of the Salmon River just upstream from where it flows along the southern edge of Cascade Head and into the Pacific.

The Oregon coast is characterized by layers of durable volcanic rock that weather to form prominent rocky hills called headlands. Sometimes these hills are isolated from the mainland by coastal erosion and become sea stacks – some of the most famous sea stacks are found near Cannon Beach to the north. The land between the headlands is usually underlain by softer sedimentary rocks that weather much more rapidly, so a drive along the Oregon coast involves a lot of ups and downs. The view of the coast from the ocean

Sea stacks on the Oregon Coast

is one of massive headlands separated by long, sandy beaches.

Most of the volcanic rock erupted sometime in the last 40 million years or so, and that gives a decent sense of the age of the interbedded sedimentary rocks as well. When liquid rock, called magma, erupts, it often releases gases that get trapped as bubbles within the magma. When the magma cools to rock, the bubbles are preserved as little round holes called vesicles. Sometimes the vesicles fill with mineral-laden water that precipitates silica in the holes, and, when translucent, this silica is.commonly called agate. As the waves wash away the headlands, they also release the agates, which get tumbled and polished in the surf where they are eagerly sought by rock hounds. Because of its long history of volcanoes, Oregon is the most agate-rich state in the nation.

In addition to the agates, beachcombers also search for round glass fishing floats that have washed across the Pacific from Japan, cool pieces of driftwood, and whatever interesting flotsam or jetsam might wash ashore.

Every once in a while, a big dead whale will wash ashore and amuse the local residents until they realize that the smell will persist and grow infinitely worse. In November of 1970, the Oregon Highway Division tried to get rid of a whale that had washed ashore in Florence by blowing it up with dynamite. But, they used too much dynamite, and the resulting blast blew large chunks of blubber through the air. Mattress-sized chunks of blubber rained down on a nearby parking lot where several cars, including those of the guys who placed the explosives, were partially crushed by falling pieces of rotten whale.

Lewis and Clark overwintered on the Oregon coast in 1805–1806 and were so starved for food that they harvested parts of stranded whale for food. Whales and other marine mammals have been washing ashore or simply dying and sinking to the seafloor for as long as there have been marine mammals, and some of these animals have become fossils that are now weathering out of the modern-day shoreline.

-EMLONG'S BESTIARY-

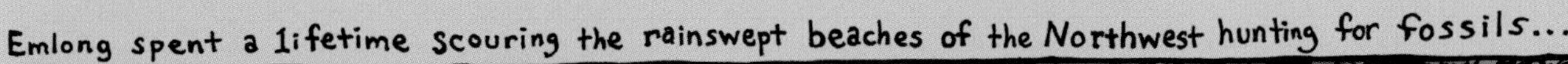

Emlong spent a lifetime scouring the rainswept beaches of the Northwest hunting for fossils...

We were finally nearing the spot where Doug Emlong had grown up and beckoned to me from the sunny pages of *National Geographic*, and I now had the opportunity to bring my childhood *Desmostylus* dreams to life. When Ray and I were in Los Angeles, we had quizzed Larry Barnes about Emlong. He had known him well, and his recollections painted a different image from the one I had constructed from my childhood memories.

Doug Emlong was born in 1942 and grew to be a curious and artistic kid in Lincoln City, where he learned early how to find agates, glass floats, and fossils on the nearby beaches. By the time he was in high school, he was finding significant marine mammal fossils. He was only twenty when he was featured in the 1961 *National Geographic.* From 1961 to 1967, he ran a small rock shop/art gallery/fossil museum called the Graveyard of the Pacific in Gleneden Beach where he sold agates, fossils, and his original paintings.

In 1968, Emlong started corresponding with Clayton Ray, the curator of fossil marine mammals at the Smithsonian's National Museum of Natural History, and the two formed a friendship. Eventually, Clayton Ray started helping Emlong pursue the science of his discoveries and began scraping money together to support his efforts.

As a collector, Emlong was relentless, scouring the Oregon coast in the worst weather for new discoveries. Eventually he traveled to Alaska, British Columbia, Washington, and California in search of fossils. With whales, pinnipeds, walrus, sea birds, desmostylians and many other marine organisms, the Emlong Collection at the Smithsonian

is truly magnificent and is still being studied today. In 2007, Daryl Domning wrote, "Douglas Emlong's Promethean prowess in discovery of unprecedented vertebrate fossils, alike in beds where many, few, or no collectors preceded him, is well known to specialists having personal knowledge of his activities." In 1980, at the age of thirty-eight, Emlong fell off of a sea cliff at Otter Crest Loop and died. The police report of the incident suggested that it was not an accident. This much I knew from reading the newspaper clippings and perusing the archives at the Smithsonian.

Barnes described Emlong as a troubled loner who had few friends and who wore magnets in his clothing to help him find fossils. Larry showed us one of Emlong's most curious discoveries, a complete and nearly perfect skull of *Kolponomos newportensis*, the oyster bear. This beast lived 20 million years ago, and Emlong had found its fossils in both Oregon and Washington. Larry described the long-fingered bear as a "beach groveler." The evolutionary history of sea lions is still being written, and one of the closest living groups to the sea lions are the bears. There is still some debate about whether Emlong's oyster bear is actually a bear.

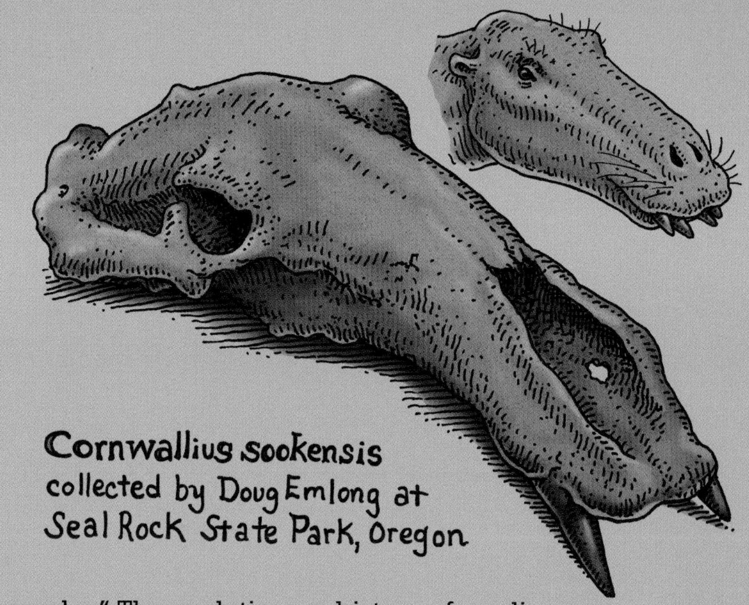

Cornwallius sookensis
collected by Doug Emlong at
Seal Rock State Park, Oregon

Frank Boyden happily displaying his spiral coprolite.

We hoped to meet people who had known Emlong so they could help us piece together the story of his short but amazing life; Ray was hoping that we could locate some of his artwork. We arrived at Frank and Jane's just in time for a perfect meal of local wild mushrooms, lightly smoked black cod from Gold Beach, and fresh king salmon from the Salmon River. Frank served it up slowly over three hours, and we oriented ourselves to our new base where we would spend the next few weeks, venturing out in all directions to sleuth out Oregon's coastal fossils. The Boydens founded the Sitka Center in 1970 as a place for scientists and artists to come together in the natural beauty of the Salmon River Valley. So here Ray and I were, coming together. Frank had met Emlong once and found him to be a bit odd and very shy with a propensity to mumble.

The next morning, we went to the Otis Café for breakfast. This snug wallet of a place only has

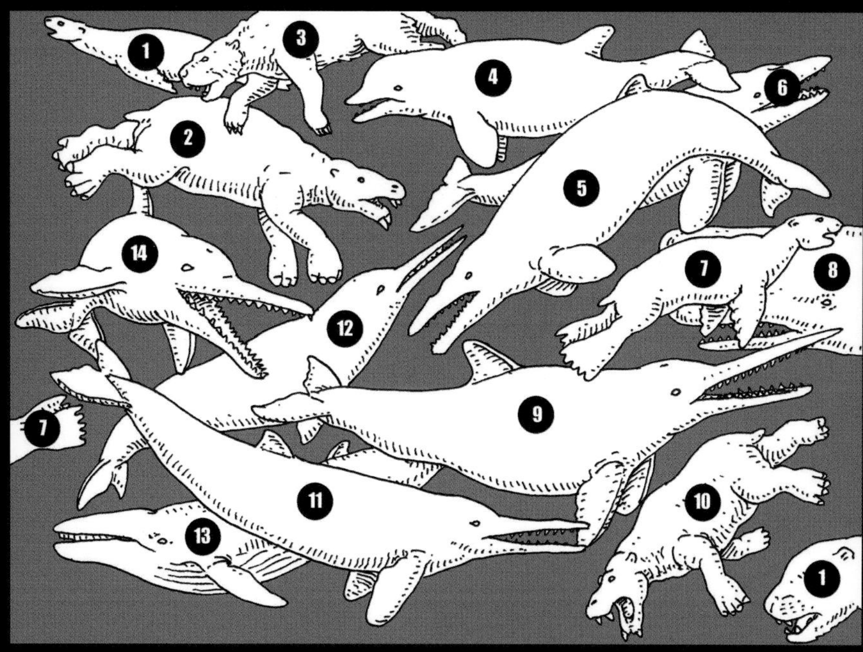


## The Fossils of Emlongia

1. *Enaliarctos emlongi* – grandfather seal
2. *Cornwallius sookensis* – early "desmo"
3. *Kolponomos newportensis* – the "oyster bear"
4. *Simocetus rayi* – pug-nosed dolphin
5. *Aetiocetus cotylalveus* – large-toothed baleen whale
6. *Fucaia buelli* – small-toothed baleen whale
7. *Desmatophoca oregonensis* – early pseudo-sea lion
8. *Physeteroidea* indet – sperm whale
9. *Macrodelphinus kelloggi* – giant long-snouted dolphin
10. *Archaeoparadoxia weltoni* – early "doxie"
11. *Llanocetus* sp. – even larger toothed baleen whale
12. *Goedertius oregonensis* – extinct long-snouted dolphin
13. *Sitsqwayk cornishorum* – early toothless baleen whale
14. Squalodontid – unnamed shark-toothed dolphin

four booths, six counter seats, and one table, but the food is amazing. They have homemade pies and black bread, either of which could kill you if you didn't have the discipline to stop eating. I nearly perished on a couple of occasions over the next few weeks. Ray struck up a conversation with the waitress and rapidly got around to asking her if she had ever met a guy named Doug Emlong. Amazingly, she had a friend who had dated him. Her memory was that he was a science nerd who was picked on in school. When the *National Geographic* article came out, he became famous – at least locally.

Frank had suggested that if Ray and I were to give a public talk about our book project, we might draw some old rock hounds out of hiding. The Sitka Center staff had found a big room at the Surftides hotel in Lincoln City and posted signs around town for a few weeks before our arrival. Interestingly, Doug Emlong's mom had once worked at the hotel's gift shop. On the appointed day, we arrived around noon and were stunned to find the room packed with more than 200 people who were eager to hear us talk about fossils. Bobby Boessenecker and his wife, Sarah, had driven up from Foster City. There was also a group of fossil collectors from an Oregon-based group called NARG (short for the perhaps too ambitious name of North American Research Group) that was founded in 2004 to support Oregon paleontology. They had recently collected a large fossil whale and were chipping it out of the matrix in one of their garages. They called it Wally the Garage Whale. The room was full of fossil-loving people, beach combers, and Ray's loyal following of T-shirt fans.

By this time, we were three years into our travels, and we were able to show lots of photos and Ray's art that illustrated what we had been

Top: A pair of desmostylid molars.

Above: Kent Gibson and his fossil marlin skull.

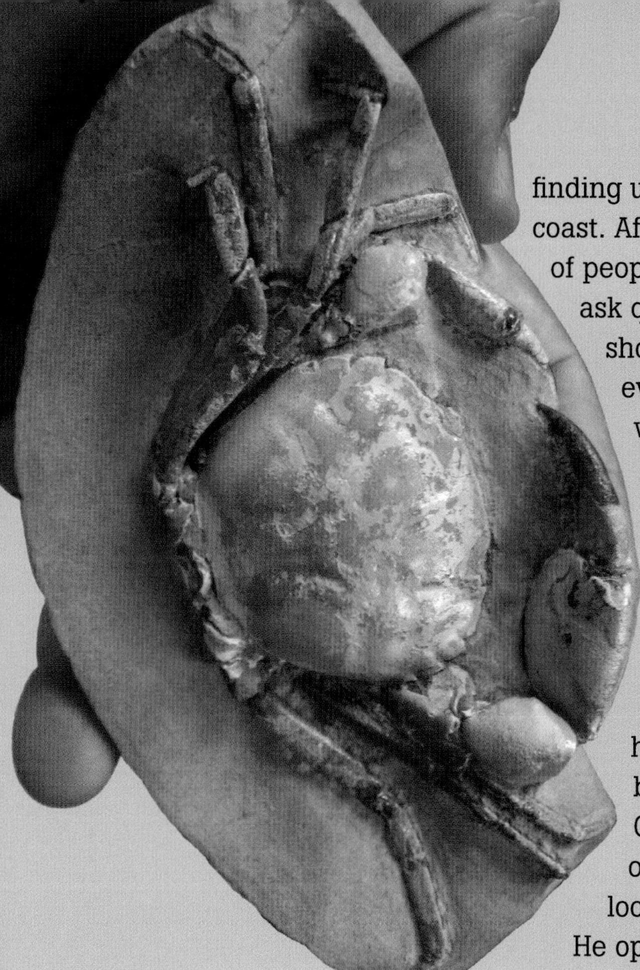

finding up and down the coast. After the talk, a lot of people stuck around to ask questions and to show us fossils. We even met a woman who had graduated from high school with Emlong.

As the crowd started to thin, a tall, tough-looking taciturn guy in a fisherman's hoody and rubber boots named Kent Gibson invited us out to his truck to look at a few rocks. He opened up the back door on his club cab and reached for some plastic buckets full of rocks. The first thing he handed me was a concretion with a very perfect skull of a sea lion–like mammal. Hello, Jane Cushing! The next rock had the back half of a whale skull. The third rock was a very beautiful juvenile *Desmostylus* skull. Hello, Doug Emlong! This was too good to be true and almost too easy. Then he said, "Wait until you see this," and dropped his tailgate to expose a rock the size of watermelon that probably weighed 45 pounds. "What do you think this is?" In a glance, I could tell that it was a skull, but it was no marine mammal. It was clearly a fish skull, and a big one at that. After some closer inspection, we convinced ourselves that it was a fossil marlin.

We invited Kent, the Boesseneckers, and the NARG guys to a nearby brewpub so we could talk fossils. I was seated across the table from a guy named Bruce Thiel, and, after the first beer, he pulled three amazing fossils out of his pocket. Each was a nearly perfect crab, complete with every single leg and claw, and each was in a

small concretion just barely larger than the crab. I had seen fossil crabs before, but I had never seen them this well prepared. Bruce collected the nodules in the creek beds of northwestern Oregon and then painstakingly exposed each crab with a tiny air scribe, a tool that is essentially a jackhammer the size of a ballpoint pen. It is extremely exacting work, and Bruce had raised it to an art form. When I stopped by his home in Portland later in the year, Bruce showed me some of the most amazing fossil crabs I had ever seen. The NARG folks were a diverse bunch, but

Above: Eocene crab *Pulalius vulgaris* from the Gary Eichorn collection.

Right: Portland-based crab-cleaning virtuoso Bruce Thiel and an exquisite *Pulalius vulgaris*.

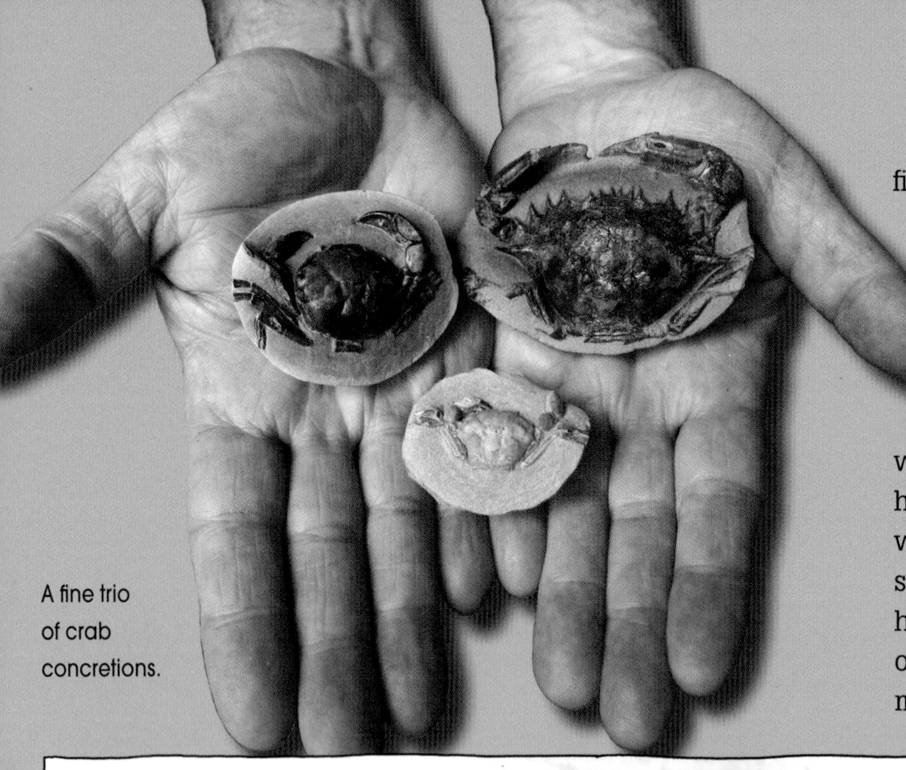

A fine trio of crab concretions.

find the spot where Doug Emlong ended his life. We were joined by Alex Krupkin, a NOAA guy who had been at the talk and who was pretty sure that he knew the spot. He brought a bottle of gin. It was around 4 p.m. in February and blustery. A brilliant sun was setting over a rough and stormy Pacific as we slowly drove up the Otter Rock Loop Road. I had seen the police report from 1980, so I knew what I was looking for and we quickly found the spot. We toasted this sad and brilliant man who had done so much in such a short time to help all of us understand the evolution of marine mammals on the Pacific Coast. Emlong left a significant legacy in all the fossils he had found and sent to the National Museum of Natural History. Living in Lincoln City, he had ready access to the entire coast, and he spent a huge amount of time exploring sea cliffs, beaches, roadcuts, and river cuts looking for fossils. He really had an amazing eye, tireless persistence, and an uncanny sense of where to look.

CULT OF THE CRAB CONCRETIONS

The next day, we drove south along the coast and made the obligatory stop at the Oregon Coast Aquarium at Newport. My family and I had visited the aquarium in 1997 to see the traveling exhibition that the California Academy of Sciences had built based on Ray's 1994 book with author Brad

they all shared our love of fossils and it was like a happy high school reunion.

After the NARG guys headed back toward Portland, the rest of us decided that we would try to

Matsen called *Planet Ocean*. The exhibition was fantastic and further convinced me that I really needed to work with Ray. Back then, Keiko the Orca, star of the *Free Willy* movie, was being held in a huge tank while waiting to be transferred to Iceland to be released back into the wild. I remember watching this magnificent beast watch me. It scared me. It also completely convinced me that whales of any size should never be in captivity.

On this visit, and because Ray is fish royalty, we got to see the back of the house, where we were grappled by an octopus and kissed by a sea lion. I noticed the beautiful cast bronze sculptures that graced the entry to the aquarium, and Ray reminded me that they were the work of Frank Boyden.

We also stopped by the Hatfield Marine Science Center, an educational facility named for Oregon senator Mark Hatfield that has been open since 1965. Every February, the Center holds a one-day Fossil Festival where local collectors display their finds and a variety of experts give lectures. I had visited a few times and had the pleasure of meeting Bill Orr who, with his wife, Elizabeth, has written a steady stream of books about the geology and paleontology of Oregon. For a state with no big natural history museum, Bill Orr has served as a one-man reference for the state's fossils.

Next, we headed south to visit a series of Doug Emlong's fossil sites. We stopped at Seal Rock State Park where he had exploited the sand-stripping power of winter storms to discover several important fossils representing four different extinct marine mammal species. He found fossils of two kinds of desmostylians: the first was the jaw of *Behemotops emlongi*, a primitive species that was named in his honor, and the second, *Cornwallius sookensis* was represented by two exquisitely complete skulls. He also found the bones of a very primitive sea lion, *Enaliarctos*. But the big find at the site was a complete skull of a whale that was very near the evolutionary split between toothed whales and baleen whales. Emlong himself wrote the first scientific description of this animal, naming it *Aetiocetus cotylalveus*.

The tide was going out when we got to Seal Rock, and the beach was covered with people walking the edge of the retreating surf in search of agates. The weather was fine, but there had been a huge storm the previous week and the beach was strewn with the carcasses of sea birds that had not survived. We found four cormorants, six rhinoceros auklets, and two tufted puffins, all reminders that life in the ocean is not easy and

Upper left: Bill Orr ponders a desmostylian snout.

that any living thing is a potential fossil. That having been said, we did not find a single fossil. Emlong really was good at finding things where everyone else could not.

We didn't find any agates either, which says another thing about the Oregon coast. A lot of people are walking these beaches and picking them over pretty thoroughly. The old-timers talk about bringing back buckets of agates. Today, you have to wake early and get out there after a big storm to find agates, and the same thing is true for fossils.

South of Florence, we stopped in at the world famous Sea Lion Caves, one of the most genuine tourist spots on the West Coast. Privately owned since 1932, this cave system in a volcanic headland is an underground sea lion rookery. It was raining pretty

**Sea Lion Cave**

One of the incredible things about the Oregon coast is the direct adjacency of living marine mammals and their fossil ancestors. The lava that made this cavern cooled about 25 million years ago, and the first true pinnipeds evolved around 26 million years ago. Geology and biology have been dancing with each other since shortly after the formation of Planet Earth. Outcrops and fossils are the memory of that long and ever-changing party.

hard when we arrived, and we entered through the gift shop and walked down a wet concrete ramp to an elevator that took us 200 feet down through the basalt headland and into the Earth. We emerged in a wet rock tunnel and walked to the underground viewpoint.

I remember this place well from my childhood as it was a mandatory stop on our way to California. The cave is large, more than 120 feet tall and quite extensive. It is open to the ocean and waves crash in with a vengeance. And the place is stuffed with sea lions, mainly the large Steller sea lion but occasionally also the smaller California sea lion. The cavern echoes with the braying barks of the 350-pound bulls. It is a mysterious and magical place.

At Coos Bay, we passed the enticingly named Fossil Point where, in 1907, Smithsonian paleontologist Frederick True discovered a partial skull of a truly massive walrus that he named *Pontolis magnus*. Emlong worked this area hard, collecting remains of a large baleen whale and four different kinds of giant extinct walrus, at least two of them still unnamed. We said our good-byes to the Boesseneckers as they headed south the California. We turned around headed back up the coast.

That night, we reunited with Kent Gibson at his home for dinner, where we met his wife, Lucy, and daughter, McKenzie. They lived up a narrow valley behind a water treatment plant and only a few miles from Moolack Beach. After breaking his back in a fall on a king crab boat in the Bering Sea in 1985, Kent decided to take a less "deadly-catch" job and started working for the Port of Newport.

Some years back, Kent bought a black Lab and named him Bart. Bart was one of those crazy black Labs that preferred to retrieve rocks rather than balls. In 1996, Kent and Bart were walking on Ona Beach, with Kent looking for agates and throwing rocks for Bart. Bart found a rock he liked and Kent kept chucking it for him. When they got back to the truck, Kent threw the rock away. Bart went and got it. Kent threw it again. Bart got it again. This went on for a while and the dog refused to get in the truck. Finally, Kent gave up and put the rock in a bucket in the bed of his truck. Satisfied, Bart hopped in the truck. When they got home, Kent took a closer look at the rock and realized that it was a fossil bone of some kind. Now he was hooked. His interest shifted from agates to fossils, and he began to be sucked into the Emlong vortex.

Kent and Lucy Gibson proudly pose with Kent's marlin skull and Ray's chalk drawing.

This was all too familiar to Lucy, whose laundry room had been invaded by concretions and fossils of all sizes and shapes. Kent is obsessed and relentless. He hits the beach first thing, every day. His proximity to the beach gives him a killer advantage over all other fossil finders. I have noticed this phenomenon in other coastal fossil sites, like the famous sea cliffs along England's Jurassic Coast. Coastal paleontology is a competitive sport.

We pored over his finds and saw a really good cross section of the types of fossils that come from Moolack Beach. There were clams, scal-

Kirk and Nick Pyenson with two fossil marlin skulls from the same beach.

lops, snails, and slipper shells, and the bones of fish, whales, dolphins, desmostylians, and pinnipeds. We took the big marlin skull out to the concrete patio, and Ray used colored chalk to flesh out the size of the whole fish. It would have been a whopper, probably in the 200- to 300-pound range. I had never seen anything like this, and I hoped that Kent would consider donating it to the Smithsonian, where it would become a scientific resource.

A few weeks later when I was back in Washington, D.C., I wandered down to the first floor of the Smithsonian's East Wing, the floor where we store many of the fossils that Clayton Ray acquired from Doug Emlong. I was looking for the *Behemotops* jaw that Doug had found at Otter Rock. As I wandered down the hall, my eye caught a familiar shape on the top shelf. I couldn't quite believe what I saw. It was a marlin skull, slightly smaller but almost identical to the one that Kent had found. I got a ladder and pulled it down. A collector named Melvin Baldwin had discovered the skull on Moolack Beach in 1964 and had given it to Doug who had given it to the Smithsonian. Two guys, two marlin skulls, and the same spot, fifty years later. I photographed Smithsonian fossil marine mammal curator Nick Pyenson holding a picture of Kent's marlin next to Doug's marlin and sent it on to Kent. This coincidence shows the importance of museums. Extremely rare finds, even ones that happen only once every fifty years, can be preserved and shared for the future. I invited Kent to come to Washington, D.C., to inspect the Emlong collection and to begin to understand how his hobby might play a larger role in the scientific understanding of the evolution of marine mammals and the history of the West Coast. Eventually, he did donate the marlin skull and the other skulls he had showed us at the Surftides.

Emlong had visited Clayton Ray in Washington, D.C., only once. He had come to see the collections and to understand what it meant that his fossils would forever be part of the National Collection.

While he was there, he slipped away one day and went fossil hunting on one of the inlets of Chesapeake Bay. Being Doug Emlong, he found a fossil whale that day.

I wanted Kent to see the famous fossil site known as Calvert Cliffs located on the Chesapeake near Solomon's Island about 50 miles from D.C., so I arranged for a friend to take him there while I was in a day of meetings. That night, Kent returned to town, and I asked him if he had found a fossil whale. He just smiled and pulled out a big perfect *Carcharocles megalodon* shark tooth out of his pocket. I can't say that I wasn't jealous.

Back with the Gibsons, Kent's daughter, McKenzie, told us during dinner that she had invented a word. Ray asked if this was something that she did often, and I asked what word she had invented. She answered, "Derpylo." We demanded to know the definition, and she told us that a derpylo is the lone banana in the grocery store. It is the one that is not part of the bunch. It's the one that is out there on its own.

Doug Emlong was a derpylo. It turns out that the sea lion Emlong displayed in the 1961 *National Geographic* wasn't a sea lion after all. When it was finally removed from its concretion, studied, and analyzed in 1995, it was given the name of *Proneotherium repenningi* (another fossil named for Chuck Repenning) and was recognized as one of the earliest members of the walrus family tree. The

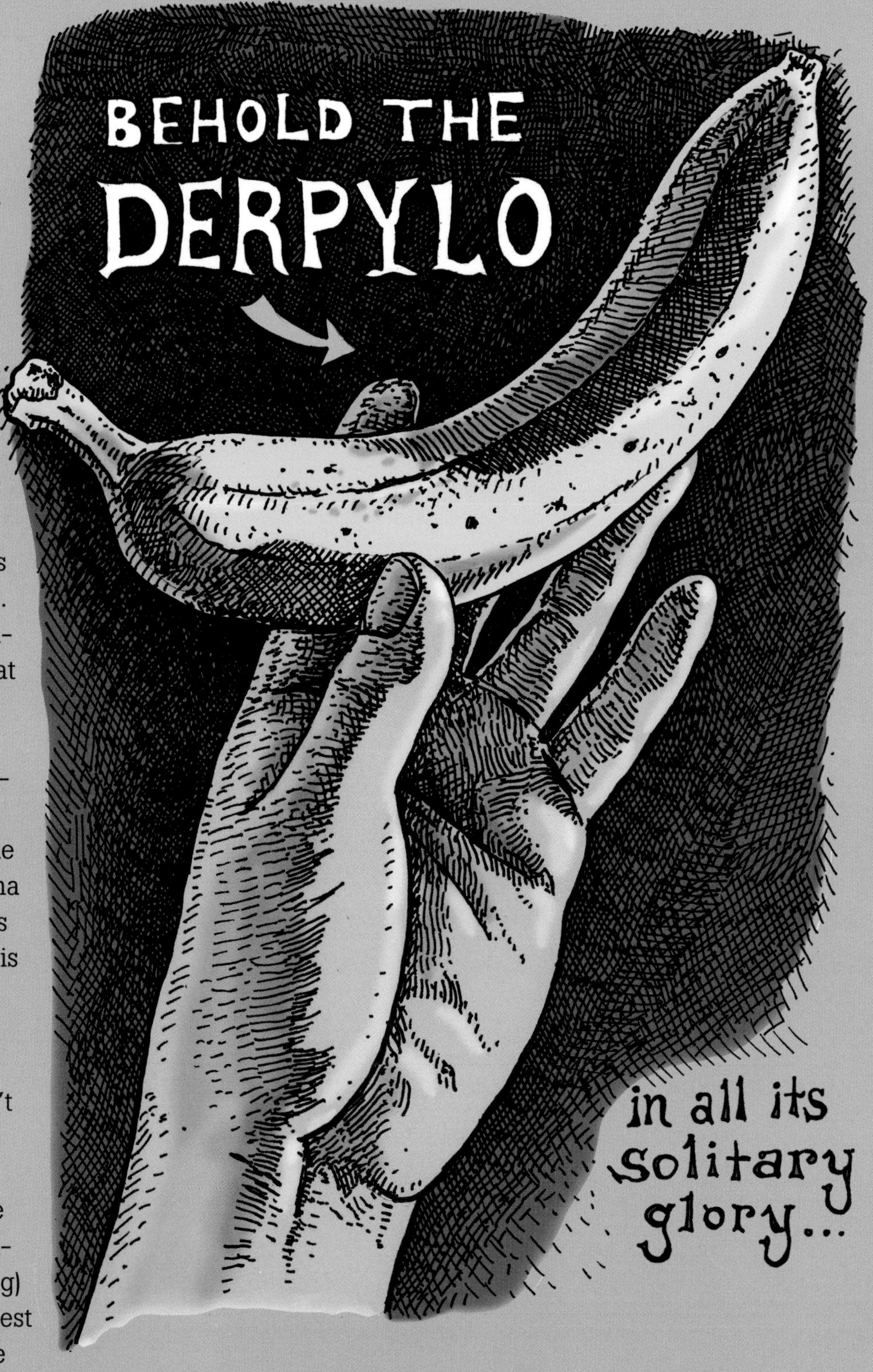
BEHOLD THE
DERPYLO
in all its
solitary
glory...

DESMOSTYLUS
PALEOPARADOXIA
OUNALASHKASTYLUS
CORNWALLIUS
NEOPARADOXIA
PALEOPARADOXIDAE
ASHOROA
DESMOSTYLIDAE
BEHEMOTOPS
HORSES
CHALICOTHERES
DESMO-
STYLIA
EQUIDAE
TAPIROMORPHS
ANTHRACOBUNIDAE
RHINOCEROSES
ANTHRACOBUNE
PHENACODUS
TAPIRS
PERISSODACTYLA
ODD-TOED UNGULATES
MAMMALIA
THE DESMO FAMILY TREE

story of the West Coast can only be told because of all the lone derpylos who have devoted their lives to their own obsessions.

A few days later, Ray and I took a break from fossiling to go steelheading. The coastal rivers of Oregon are known for more than their fossils. Frank Boyden had become friends with a tough young fishing guide named Silas Stardance. The child of hippies, Silas was finding his way through fish. We took his drift boat up the Siletz River to a spot where it was only 40 feet wide and launched into the swiftly moving green water. Being February, it was cold and damp. In a boat with two artists, one paleontologist, and a guide, we found ourselves watching the banks as much as the pools and riffles. A few times we stopped to inspect outcrops or fallen blocks of rock. We had Emlong on our mind.

The Siletz was the river that Ken Kesey wrote about in his novel *Sometimes a Great Notion*. When I first read that book as a teenager, I was struck by the harsh realities of being a logger in a coastal rain forest. If you've read the book, you'll recall that it doesn't end well, either for the forests or for the loggers. Frank remembered when Paul Newman came to the Siletz to make the 1970 film and how the town of Lincoln City was starstruck.

Just before noon, I hooked into a sea-bright steelhead and eventually fought it into the boat. There is no fish more beautiful than the steelhead trout. Like their salmon cousins, the steelhead spend their adult lives in the Pacific Ocean as full players of the open ocean ecosystem. When the time comes for them to reproduce, they swim from the ocean into coastal rivers, switching from saltwater to fresh water to spawn. The evolution of steelhead and salmon was one of Ray's many obsessions, and we were happy to be on a river in steelhead country.

That fish turned out to be a derpylo. After six hours of drifting, we were all getting cold and tired. Silas rowed the boat to shore so Frank could relieve himself. I sat in the boat, scanning the cliff banks for signs of fossils. You can't find them if you don't look for them, and the funny thing about looking for them is that you often forget that you are looking, and you are surprised when you make a find. My lazy scan caught a glimpse of white. I paused and then recognized an unmistakable pattern of an occipital condyle, the base of a skull sticking out of the rock of the riverbank. I couldn't believe it. Here we were, floating down Paul Newman's river with a steelhead in the boat and there was a fossil skull.

I yelled and jumped out of the boat and ran to the bank surging with anticipation. Just before I got to the fossil, I froze and turned back toward Frank. His huge eyebrows could not conceal his twinkle, and I realized that I had been had. I bent down and pulled the skull out of the bank. It was a beautiful complete skull. But it was made of porcelain and engraved with the telltale marks of a certain coastal ceramicist named Frank Boyden.

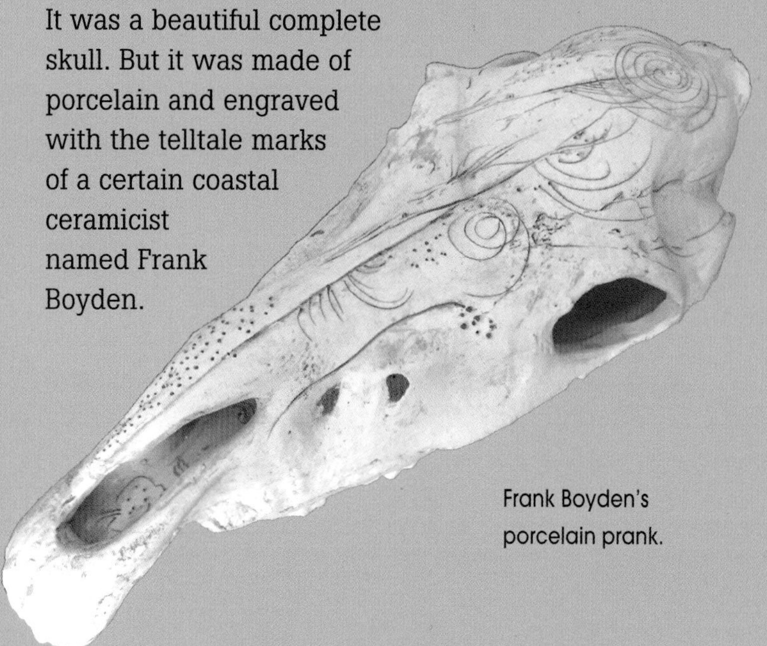

Frank Boyden's porcelain prank.

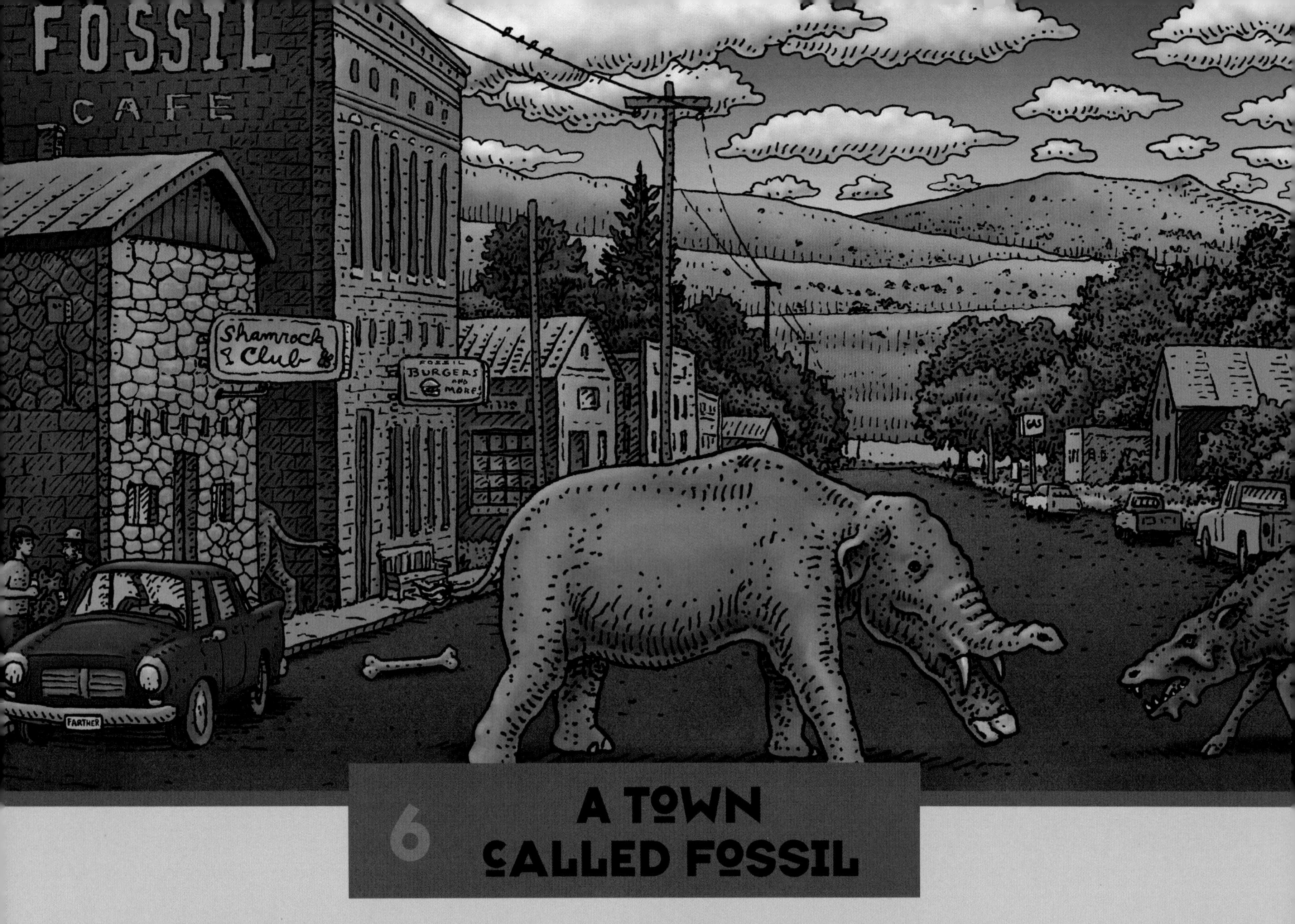


# 6 A TOWN CALLED FOSSIL

The Oregon state animal is the beaver, its state fish is the Chinook salmon, its state rock is the thunder egg, and its state fossil is the *Metasequoia*, a fact that delights me but frustrates animal lovers like Ray. When Ray and I ventured away from the coast to explore the rest of the state, our first stop was the University of Oregon in Eugene, home of the Ducks. Just up the road in Corvallis at Oregon State University are their fearsome rivals, the Beavers. For a state without a big natural history museum, natural history is oozing out everywhere.

My memories of Eugene were painful, since the only significant time I'd ever spent there was to have a knee surgery in 1979. I remember looking out from the University Hospital to see the fraternity row where John Belushi's 1978 film, *Animal House*, was filmed. Eugene shows its roots as a hippie town with a bronze statue of Ken Kesey, the Merry Prankster. And there are ducks, lots of ducks.

A shovel-tusked *Amebelodon* faces down an *Archaeotherium* "killer pig" on a dusty street in Fossil, Oregon.

This time, our goal was the University of Oregon Museum of Natural History and Cultural History (though many people just call it the Condon Museum). Until 1936, the museum was named for Thomas Condon, an Irish preacher who moved to Oregon's Willamette Valley in 1852. He was interested in rocks and fossils, read scientific journals, and began to correspond with famous paleontologists. In 1862, he moved to the rough-and-tumble town of The Dalles, a great place for a man who could be called Oregon's first true rock hound. In 1865, he accompanied a group of soldiers into the remote center of Oregon, now known as the John Day Basin, and realized that the region had tremendous potential for fossils. In 1869, he was enticed to ship a load of fossils to the Smithsonian Institution. In 1871, Yale's O. C. Marsh arrived with a dozen undergraduates and was guided to the fossil beds by Condon. Later that decade, Edward Cope, Marsh's archrival from Philadelphia, sent his collectors to the John Day country, and the eastern Oregon fossil rush was on. By 1872, Condon's reputation was so good that he was named as the young state's first state geologist, and in 1876, he became one of the first six faculty members of the newly formed University of Oregon. By 1900, the remote John Day Fossil Beds were known to the paleontologists of the world, and Condon had a global reputation.

In 1902, Condon wrote a book entitled *Two Islands and What Came of Them*. This was a deep-time tale of the fossil history of eastern Oregon that showed Condon's deep insights into the processes of geology and paleontology and the

Far left: Thomas Condon, the original Oregon fossi hound.

Left: University of Oregon students John Orcutt, Kelsey Stilson, Nick Famoso, and Amy Atwater with the skull of a giant spike-toothed salmon.

recognition that ancient Oregon was wholly different from the Oregon of today. In 1906, he found and described a fossil seal (*Desmatophoca oregonensis*) from the Oregon coast, setting the stage for Doug Emlong a half a century later.

Condon's legacy in paleontology is vast, but his footprint on the campus today is modest. The small museum sits on a neighborhood street near the university, and many of the collections are stored in a basement room on campus. Three paleontologists, Greg Retallack, Samantha Hopkins, and Ed Davis, work at the university, and they were all there when Ray and I rolled into town on a rainy afternoon.

Greg is a paleobotanist whom I first met in 1983. I understood him to be an eccentric Australian with very broad interests and one very deep passion. Over the years, Greg had become obsessed with fossil soil and had come to be known as the "man that reads dirt." For Greg, there is no problem in geology that cannot be solved with soil. I was dying to see what would happen when I put him and Ray in the same room.

That room was Greg's office in the geology building, where he showed us a surprisingly comprehensive collection of world fossils tightly curated into a few cabinets in his office. He called it his "aesthetic indulgence." Ray was used to meeting paleontologists who focused on specific types of organisms or certain time periods, and he was baffled by Greg's encyclopedic view of ancient worlds. Every time Ray mentioned a fossil or theory, Greg would spin on his chair, reach into a drawer, produce a paper, and state something like, "Here is my theory on that topic." Remarkably, almost every single one of his theories was strongly at odds with the prevailing wisdom of the field. When I pushed him on this, he said, "You win a few, you lose a few, mainly you lose a few." He is a committed paleo-contrarian. He is also quite generous and invited us to dinner and to spend the night at his home.

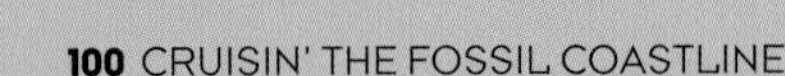

Later that night over a dinner of steak and beans, we chatted about giant zebra-like Pliocene horses from Idaho and the world's largest dog, a Nebraska monster known as *Epicyon*, typical fare when paleonerds gather.

The next morning, we made our way back to campus and met Samantha Hopkins; her husband, Ed Davis; and four of their students. They were all working on various Miocene fossils, and the room was abuzz with the energy of new discoveries. Samantha told us that she was the child of hippies and that she rebelled by not smoking pot and by becoming a scientist. One of her students, Amy Atwater, a graduate of Camp Hancock, writes a blog called *Mary Anning's Revenge*.

And these guys had amazing fossils. They showed us the upper arm of an unnamed fossil cat from 7-million-year-old rocks near The Dalles

## Chasing the Giant Salmon

Ray had been on the trail of this giant salmon, *Oncorhynchus rastrosus*, for two decades and had once visited Eugene to make sure that the stories were true. During that visit, he had talked to Bill Orr, then curator of the UO museum, into allowing him to take the actual type specimen of the fossil out of the museum and to lay it out on a sidewalk so that Ray could chalk out the rest of the animal and get a better sense of how huge it really was. Pushing 8 feet and more than 500 pounds, it was an absolute monster, or a whole lot of lox, depending on your perspective. In an exhibit at the Denver Museum of Nature & and Science in 1999, we had mounted a fleshed out giant spike-toothed salmon and mounted it next to a saber-toothed *Smilodon*, and it looked like a pretty fair fight.

## A Salmon by Any Other Name...

Ed Davis had recently returned to the salmon's discovery site near Madras and realized that there were more bones to be found, so he has started a new excavation. It turned out that, after our visit, Ed CT-scanned the salmon specimens they had and started to realize that the tusks didn't project downward as they had originally thought but instead more toward the side. This caused a significant amount of heartburn to Ray who had invested many hours drawing this toothy mega-fish, and it troubled him deeply that a saber-toothed salmon might have to be renamed, and in fact redrawn, as a buck-toothed, side-toothed, or slack-toothed salmon. In the end, they settled on the name of spike-toothed salmon.

that would have been significantly larger than an African lion. We wandered the cramped collection room with Ed and saw amazing treasures: mammoths from many places, a giant sloth from Hillsboro, yet another marlin skull from Moolack Beach, ammonites from Mitchell, and best of all, the remains of a giant spike-toothed salmon from Madras.

Ed Davis and the femur of a giant ground sloth.

We said our farewells and headed east with the idea that we might be able to visit the whatever-toothed salmon fossil site near Madras. Driving out of Eugene, we took Highway 126 over the Cascades, past the Three Sisters Mountains, and down into Prineville. This part of central Oregon is a mecca for rock hounds because of a curious type of rock known as a thunder egg. Most of these bumpy round rocks are about the size of a baseball, but they can get a lot larger or a little smaller. They are prized because they are often filled with agate that formed in pleasing patterns. To realize the potential of a thunder egg, you must first cut it in half with a diamond-bladed saw to expose the configuration of the agate within. Then you use increasingly fine grits to polish the sawn face. It's a fair bit of work for a modest reward, but bands of rock hounds from around the world trek to central Oregon to dig thunder eggs. Of course, you can also buy them in a cut or uncut state. I had never paid too much attention to them because they were not fossils, but lately, I have had a change of heart.

A pile of uncut thunder eggs and four examples of what happens when you cut them open.

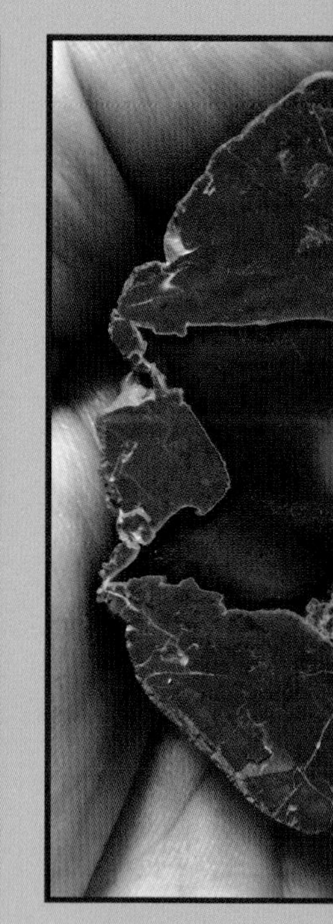

Oregon, which seems to be a state with limitless mascots, formally adopted the thunder egg as its official state rock in 1965. I imagine that this was near the peak of the thunder egg frenzy, and the frenzy has definitely calmed down. The best way to experience them these days is to simply type the words into an online auction site, and you'll see the curious little world of thunder eggs on display at very modest prices.

We stopped at a classic rock shop called Richardson's Rock Ranch in Madras so I could initiate Ray into the clan of the thunder egg. The place was great. It had a huge showroom full of sliced and polished thunder eggs and other rocks, and a huge yard full of "rough" that was for sale by the pound. Peddling rocks is a family business, and the whole Richardson clan was sitting in the showroom cracking jokes and having good time.

The Richardson Rock Ranch would remind you of a fruit stand where you can buy over the counter or go pick your own. At the rock ranch, you pay a token fee and trundle off to the digs with a shovel and crowbar to dig your own thunder eggs.

How the agate gets into the thunder egg is a story about the geology of eastern Oregon, a place that has been venting gases, erupting lava, and exploding ash for a very long time. The state has a long and complicated volcanic history, and it is this history that explains why both fossils and thunder eggs are found in abundance.

Volcanic eruptions are quite variable, both in the composition of what is erupted and also how it erupts. In general, the more silica that is in the molten rock, the more explosive the eruption. Mount St. Helens exploded in 1980 in part because the chamber beneath it contained a magma that was rich in silica. It is this sort of eruption that creates airborne volcanic ash that rains out of the sky and buries landscapes.

This is a great way to make fossils because the mechanism both kills and buries plants and animals. The ashfalls that happen close to the volcanoes often come down hot and weld to-

gether into a rock almost immediately. This is what I call airborne death rock. Geologists call it a welded tuff. Just imagine if the sky above you started raining red-hot concrete.

Farther from the eruption, the ash has a chance to cool, and it rains down as if there were a great sifter sprinkling sugar and flour out of the sky. In both cases, the results can be devastating, and the really big eruptions can bury landscapes under dozens of feet of welded tuff or ash.

Eruptions of silica-rich lava can also form rivers of molten rock that flow across the landscape, filling valleys and creating new topography. There are places in eastern Oregon that are composed of pure glass called obsidian that formed in this way. Obsidian is perfect for making sharp tools and arrowheads, and Oregon obsidian artifacts show up all over the American West.

Thunder eggs are found in the weathered remains of silica-rich lava flows. This tells us that they must have been formed by a process that started with a lava flow. Geologists speculate that they form when one of these hot lava flows entrain gas pockets that leave voids as the lava cools. Later, silica-rich waters percolate through the ash, depositing both concentric and layered bands of colored silica. The process also seems to involve a phase of expansion and collapse as the mineralizing fluids flex the constraining bedrock. In the end, the result is knobby eggs, that when cut, show patterns that are varied and beautiful, with some of them looking like landscapes, clouds, or animals.

Eventually the lava rots to a rubbly clay, and the "U-pick" diggers at the rock ranch can root out the thunder eggs like they were digging potatoes.

The volcanoes of central Oregon were blowing and going for the last 45 million years, and the result is an incredible landscape of weathered lava flows, ash beds, and sedimentary rock. We had squandered our chance to hunt for spike-toothed salmon by spending so much time at the rock ranch, so we left the giant fish behind and headed into John Day Country.

John Day was a fur trapper from the East Coast hired to join John Jacob Astor's company and ended up in Astoria in 1810, less than a decade after Lewis and Clark. While trapping along a tributary of the Columbia River, he encountered a band of Native Americans that robbed him of his furs and all of his clothes. He was forced to make his way back to the Columbia River buck naked. That river now bears his name, and the landscape it carved is one of the most prolific and storied in North America. We were headed to the John Day Basin and the widely dispersed units of the John Day Fossil Beds National Monument.

The roads in eastern Oregon are long and deserted. Highway 26 from Prineville to Dayville is a beautiful drive through the pine forests of the Ochoco National Forest. We stopped in the tiny town of Mitchell and took a look at the resident bear named Henry who lived in a big garage-sized log cage in the middle of town. The nearby hills contain outcrops of Cretaceous marine mudstone that yield ammonites and marine reptiles. One local plesiosaur skeleton has been called the Mitchell Monster or the Tiger of the Seas.

Hemipsalodon skips the salad bar yet again.

The Painted Hills.

Mitchell calls itself the gateway to the Painted Hills, and that was why we were there. A winding 9-mile drive later and we were in a valley of rounded muddy hills that presented a wildly surreal palette of the grays, greens, and reds of the Clarno Formation. This is one of those places that defies description, as its beauty shifts throughout the day with the angle of the sun and the amount of moisture in the ground. Ray was stunned by the unexpected polychrome vista and wondered out loud about what could cause the brilliantly colored stripes in the hill. We had left the answer back in Eugene, as the stripes are cross sections of ancient subtropical forest soils. This is the kind of place that vindicates Greg Retallack's lifelong obsession with fossil soil.

It was also the first fossil site that Thomas Condon had visited in his inaugural trip to central Oregon in 1865, and it was here in 1897 that J. C. Merriam collected the specimen that came to be known as the Berkeley entelodont, an animal that Ray calls a "hell pig." Condon also found fossil leaves in the nearby hills.

There is no better place to see what we know about the lost worlds of John Day than the Thomas Condon Paleontological Center in Dayville, Oregon. Built in 2003, this beautiful modern museum is owned and run by the National Park Service. The museum is a very long way from anywhere, and it is only the lucky or determined visitor that makes it to its doors, but it is certainly worth the drive. Shortly after the Park Service took over the management of the many John Day fossil sites from the state of Oregon in 1975, it hired a young paleontologist named Ted Fremd. Ted spent his entire career discovering and collecting mammal fossils from old and new John Day sites. He also visited museums around the country that had fossils that had been collected since the time of Condon. While he was doing this, he was also sorting out the geology of the basin and understanding the relationships of all the various fossil-bearing layers. The early paleontologists had been aware that there were fossils from many different time periods, but Ted was able take advantage of new dating tech-

## An Asian Forest on North American Soil

**In the 1920s, Ralph Chaney from UC–Berkeley discovered a fossil plant site near the Painted Hills that yielded fossil leaves of alder, oak, walnut, dawn redwood (*Metasequoia*), and Japanese Katsura tree (*Cercidiphyllum*). It would be a few years before he realized that his 31-million-year-old fossil leaves were related to the forests of Asia, but he could certainly tell that what was a desert today had once been a dense forest long ago. He called his discovery the Bridge Creek Flora and went on to compare it with the ecology of the modern coastal redwood forests. He was one of the first paleobotanists to use counts of fossil leaves to compare with modern leaf litter and reconstruct ancient forests. The Bridge Creek flora was just one of his discoveries and, working alongside Merriam and other UC–Berkeley paleontologists, they began to reconstruct a whole series of ancient worlds.**

niques that allowed him to put actual numerical dates on the fossil sites and to begin to calibrate the sequence of fossils. Tiny crystals of sanidine and zircon in the volcanic ash contain minute amounts of radioactive elements that allow the ash beds to be precisely dated. Ted describes the volcanic ash beds as "the page numbers in the book of the Earth."

Both plant and vertebrate fossils are abundant here, and the stacked layers of sedimentary rock interspersed with lava flows and ash beds made the perfect place to tell a story of evolution and of Oregon's changing ancient landscapes. The stack of layered formations is hundreds of feet thick and preserves a 37-million-year story of evolution. Working with paleobotanist Regan Dunn and a team of National Park exhibit designers, Ted created a museum that tells this story with amazing grace and clarity by presenting seven different ancient worlds, each represented by a beautiful painted mural and a suite of fossils from that site, including the 44-million-year-old (mya) Clarno Assemblage, the 40-mya Hancock Quarry, the 33-mya Bridge Creek Assemblage, the 29-mya Turtle Cove Assemblage, the 24-to-20-mya Upper John Day Assemblage, the 15-mya Mascall Assemblage, and the 7 mya Rattlesnake Assemblage.

We met museum curator Josh X. Samuels who invited us into the fossil lab where technicians were busily extracting small skulls from a light-gray matrix. The adjacent collection room was full of superbly preserved bones, teeth, and leaves. A tally of fossil sites from the monument lists nearly 800 individual quarries. Between the animals and plants, the sites and ash beds, John Day is one of the clearest fossil views of evolution in the world.

Ted Fremd and a jawbone of *Archaeotherium*, the killer pig.

POLAR BEAR
BROWN BEARS
SUN BEAR
SLOTH BEAR
ANDEAN BEAR
SHORT FACED BEAR
WALRUSES
SEA LIONS
SEALS
ASIATIC BLACK BEAR
BLACK BEAR
URSIDAE
BEARS
PINNIPEDS
PANDAS
OYSTER BEAR
SKUNKS
MUSTELIDS
PROCYONIDAE
CANIFORMIA
"DOG-LIKE"
BEAR DOGS
AMPHICYONIDAE
CANIDAE
WOLVES
DOGS
HYENAS
RACCOONS
OTTERS
WEASELS
CATS
CARNIVORA
MAMMALIA
THE BEAR FAMILY TREE

We got sucked into the exhibition and to the worlds it portrayed. Josh guided us through the stack of formations and their extinct faunas and told us amazing things about animals that we had never heard of before. The Turtle Cove Assemblage alone had ten different kinds of fossil dogs and five different types of saber-toothed cats. The first beaver appeared around 7 million years ago, launching Oregon's ongoing rivalry with the duck. *Hemipsalodon* was a grizzly bear–sized creodont, *Enhydrocyon* was a "hypercarnivorous" dog, *Dromomeryx* was an antelope-like animal with goal post–shaped horns, *Tephrocyon* was the first of the bone-cracking dogs, *Indarctos* was a bear as big as a polar bear, and *Cynarctoides* was a Chihuahua-sized big-eyed nocturnal arboreal dog. The concept of a tree-climbing dog had never occurred to me, so I made the unfortunate pun about the "tree-dog night." This was clearly a mistake that Ray could not resist.

With all of its fossil dogs and bears, the John Day region is really a place to ponder evolution and to stand back and admire the family tree of bears, dogs, extinct bear-dogs, and their relatives the seals, sea lions, and walrus.

Just a little way up the road we stopped for a hike up into a valley called Blue Basin, which had produced fossils of the Turtle Cove Assemblage. It was here in 1878 that a twenty-eight-year-old Charles Sternberg, one of Edward Cope's collectors, had found the skull of *Pogonodon*, a saber-toothed cat, perched high on the top of a rock spire called a hoodoo. In one of the great collecting stories of all time, he describes how he shimmied up the spire to grab the prize on the top before it all came crashing to the valley floor.

We spent that night in the town of John Day. The next morning at the local café, we met a heavy equipment operator named Moose. It is always worth interrogating equipment operators because it is they who are truly on the edge of the blade all day, and they will be the ones to find and, all too often, bury fossils. Moose was quiet, but finally he did allow that he had seen some of those "big curly things" down by Seneca.

Suspecting that we knew what he meant, we took the 30-mile drive to Seneca and found ourselves in a big rifle range dug into the side of a hill. Eastern Oregon is known for being well-armed, and the rocky rubble of the range was littered with bullet casings, shotgun shells, and shattered clay pigeons. The rock looked shaley and promising in a fossily sort of way, but we couldn't find anything at first. When it comes to fossil hunting, though, patience is a virtue, and we quietly split up and slowly wandered around the pit. After forty minutes of big nothing, we decided to give up and head back to the John Day Fossil Beds, and just at that moment, I noticed a curly thing. There amongst the clay pigeons was a Jurassic ammonite. We had found the sea that surrounded one of Condon's lost islands of Oregon. Thank you, Moose.

The next stop was the town named Fossil. I had been there before, and I will go there again. The place is magical in a very peculiar and remote sort of way. When I was a fossil kid, it drove me crazy to look at the map of Oregon and see the town of Fossil. I made it there shortly after getting my driver's license and was delighted to learn that the bank on the back side of the high school's football field was full of fossil leaves. It was also great to learn that the United Bank of Oregon had a Fossil Branch and that it was possible to engage in Fossil Bible Study, eat Fossil Burgers, buy Fossil Auto parts, and attend Fossil School. My Nirvana would look like this place.

When we had stopped at the Condon Museum to see Ted Fremd, he warned us that the one gas station in Fossil only sold gas from Wednesday through Saturday and only between 2:30 and 5:30 p.m. We arrived in Fossil just in the nick of time to fill our nearly empty tank.

We pushed on toward Clarno to visit Camp Hancock, the Oregon Museum of Science and Industry paleontology camp that had been established by Lon Hancock. It was here that he had made his big discovery that would come to be known as the Hancock Quarry. Despite all of the amazing work that Condon and the people who followed him had done, they had mainly found Oligocene and Miocene fossils. In the hills just above the camp, Hancock hit the jackpot with a 44-million-

Josh X. Samuels and the skull of a tiny arboreal "tree dog."

year-old site and the first Eocene mammal fossils of eastern Oregon. In actual fact, the first bone was found by one of Lon's friends but he led the subsequent excavations. For years, teams of teenagers from Portland would trek out to Camp Hancock and labor away in Lon's quarry. The results were staggering. Just a reading of the names of the extinct animals of the Hancock Quarry was like reading a lost Greek poem:

***Hemipsalodon, Eubrontops, Zaisanamynodon, Protitanops, Prototapirus, Halohippus,*** **and** ***Epihippus.***

**(In English, these beasts were rhinos, tapirs, titanotheres, creodonts, and horses.)**

On a nearby hill in slightly older rocks, Lon's friend, a printer named Tom Bones, toiled away on a different project. Despite his name, he was not a bone digger. He was after fossil nuts. The Clarno Nut Beds have come to be known as one of the great fossil plant sites on the planet. In what was an ancient mudflow, twigs, nuts, fruits, and seeds are preserved in near perfect petrifaction. Bones started chipping away at this cliff back in the 1940s, and in a 1977 newspaper article he was quoted as saying, "Nobody seems to know anything about it so it seemed like a good thing to do." In 1961, Bones donated his collection to the Smithsonian and startled the paleobotanical world with what he had found.

And Bones's legacy extended beyond his collection of fossil plants. Three kids, Jack Wolfe, Steve Manchester, and Herb Meyer, who had worked with Tom Bones as students at Camp Hancock, went on to become renowned paleobotanists and all would make contributions to understanding the paleobotany of John Day. Jack Wolfe used his teenage fossil skills to win the national Westinghouse Talent Search, which launched him to Harvard and on to the US Geological Survey where he became the leading figure in West Coast paleobotany. Steve Manchester, now a paleobotanist at the University of Florida, exhaustively studied the Clarno Nut Beds and described more than 173 plant species from

A petrified walnut (*Juglans clarnensis*) from the Clarno Nut Beds.

Kirk, age seventeen, visiting Tom Bones in 1977.

what was clearly a tropical rain forest. The site has even produced fossil banana seeds. Herb Meyer described the flora from the Painted Hills and now works for the National Park Service at the Florissant Fossil Beds National Monument.

On my first trip to hunt for fossils in eastern Oregon in 1977, I stopped in Vancouver, Washington, to meet Tom Bones. He was eighty-five years old at the time and delighted that his nuts had amounted to something. He was once quoted as saying that of the countless hours he had spent collecting fossils over an interval of more than forty years, almost all of them had been spent within a 50-foot radius.

On that trip, I had also stopped at the Oregon Museum of Science and Industry (OMSI), where Steve Manchester was working with a group of a dozen high school students who had collected thousands of fragments of mudstone that had leaf imprints. They were gluing the pieces together with Elmer's Glue and reassembling complete large leaves that had the characteristic oval shape and elongate tip that characterizes modern tropical rain forest leaves. At the time, I thought it was the most insane thing I had ever seen,

## "Property of UCMP"

In 1978, a friend of mine went to a garage sale in Portland and spotted a couple of cardboard boxes full of fossils. He bought them for me knowing that I would pay him back. As I dug into the boxes, I was surprised to see fossil jaws with museum numbers on them. On a few of the specimens, I could make out the letters "UCMP." Even then, I could recognize the initials for the University of California Museum of Paleontology. I mailed the fossils to Berkeley only to learn that they were specimens that had gone missing from their collection. I'll never know for sure, but I think that I had stumbled on remains of the Hancock collection that had not made it to the OMSI collection.

but years later, I visited Steve in his lab at the University of Florida and saw the results of their labor. Between the leaves, twigs, nuts, fruits, and seeds, Steve had assembled the richest single fossil plant site known in the world.

The John Day beds are remote and little known, yet they have seen a steady stream of paleontologists since the 1860s and have produced one of the greatest fossil records of the world. Much of the work was done by professionals, but Tom Bones and Lon Hancock, a printer and mailman, played critical roles.

Leaving Fossil, Ray and I drove north to the Columbia River and crossed it near Maryhill, Washington, where, in 1918, a curious fellow named Samuel Hill had constructed a concrete model of Stonehenge as a World War I memorial. We followed the river downstream toward Portland. The river cut through thick layers of Miocene basalt, and Mt. Hood loomed ominously over the town of Hood River. Ever since the 1980 eruption of Mount St. Helens, the volcanoes of Washington and Oregon have demanded more

TWO WANDERING DAEODONS ENCOUNTER A VERY ANNOYED TYLOCEPHALONYX SOMEWHERE IN ANCIENT OREGON FIFTEEN MILLION YEARS AGO...

respect for the hazards that they are. Eruption of molten rock has been a continual theme for the last 40 million years, and many of the fossils of Oregon owe their presence to the fact that this part of the West Coast has very dynamic, explosive, and eruptive geology.

The gorge of the Columbia is where the river has cut its way through the Cascade Mountain Range on the way to the Pacific Ocean, and it was here that the Bonneville Dam was completed in 1938. Thirteen more dams would follow, creating irrigation and cheap energy but dooming the mighty runs of Chinook salmon.

Oregon's Mount Hood towers above the basalt layers of the Columbia River Gorge.

We drove along the Washington side of the river, and in Longview, we crossed the Cowlitz River where it flows into the Columbia. When Mount St. Helens erupted, the mudflow that roared down the Cowlitz was strong enough to carry bridges away. We think of volcanoes as erupting lava, but in the Pacific Northwest, the glaciated slopes of the big volcanoes are charged with ice and mud that can be incredibly destructive in their own right. Many of the valleys that radiate away from the mountains now have mudflow evacuation signs, an attempt to prevent people from becoming modern-day fossils.

We were listening to the local NPR station as we drove west along the shore of the Columbia River approaching Astoria, and the newscaster mentioned that a pair of humpback whales had crossed the bar and were headed upriver. I looked left and, incredibly, there were the whales. We were back on the coast.

We crossed the river back into Oregon and headed down the coast past Cannon Beach where the massive sea stacks attract tourists, but the tsunamis signs suggest they should be careful where they sleep. That evening, we were back on the Salmon River with Frank and Jane Boyden, regaling them with fossil tales from the eastern part of Oregon.

WASHINGTON
STATE FOSSILS
Vancouver Island
Nanaimo
Vancouver
Delta
Ammonites
Duncan
Kolponomos-the "Oyster Bear"
THE SUPER STRANGE Behemotops
Victoria
Dinosaur Bone!
W.W.U. GEOLOGY DISPLAYS
Ferndale
San Juan Islands
Bellingham
Mount Baker
CRETACEOUS AMMONITES
NORTH CASCADES SCENIC HIGHWAY
CARBONIFEROUS CRINOIDS
HARTS PASS
Oroville
Stonerose Interpretive Center Eocene Fossil Site
EOSALMO
Tonasket
Republic
Insects too!
CRETACEOUS FERNS AND CYCADS
Winthrop
Twisp
Omak
Okanogan
Rockport
Mt. Vernon
Anacortes
OAK HARBOR
Coupeville
Camano
Neah Bay
LONG BEAKED DOLPHIN
CRAB NODULES
Port Townsend
PALM FRONDS CHUCKANUT FORMATION
CAN YOU DIG IT?
Port Angeles
Sequim
WORLD FAMOUS MANIS MASTODON SITE
Mammoth Tusk from Downtown Seattle
"Lulu"
Brewster
Bridgeport
Forks
La Push
Hoh River
Mt. Olympus
Everett
BURKE MUSEUM of Natural History + Culture
DAWGS
Lake Chelan
Chelan
GRAND COULEE
Grand Coulee Dam
Electric City
Columbia River
Ruby Beach
Columbian Mammoth
Pacific Science Center
Bellevue
Dentalium shells
Bacon
Fossil Sea Star
Bremerton
Port Orchard
Seattle
Issaquah
Leavenworth
Entiat
THE AMAZING RHINO-SHAPED CAVE OF BLUE LAKE, FORMED WHEN HOT LAVA ENGULFED A DICERATHERIUM.
Coulee City
Wenatchee
Soap Lake
CAMEL
Vashon
GIG HBR
Fossil Sockeye
Maple Valley
Quincy
Taholah
Aloha
Humptulips
MATING Slipper Shells
Tacoma
Auburn
Black Diamond Coal
DIATRYMA FOOTPRINT OR...?
GINKGO PETRIFIED FOREST STATE PARK
BILL ROSE THE PALEO LUMBERJACK
Ephrata (nuthin!)
RING-TAILED CAT
THE SEATAC SLOTH
Enumclaw
Ocean Shores
Aberdeen
Olympia
Lacey
Buckley
Ellensburg
Moses Lake
Ritzville
DINOFELI
Ancient Baleen Whale-on display at the Burke!
Westport
THE MIMA MOUNDS OF MYSTERY
Little Rock
Centralia
Mt. Rainier National Park
FOSSILIZED NUT HOARD
Vantage
PALEO RED PANDA
Raymond
South Bend
OYSTER CAPITAL OF THE WORLD
Chehalis
Have no fear it's a Gomphothere
Naches
Yakima
Aturia Nautiloid
Pseudo-Coprolites
Castle Rock
Mount St. Helens
Sunnyside
Spike Toothed Salmon!
GROUND SLOTH
Longview
D.B. Cooper's
LOOT
Kennewick Man
Richland
Pasco
Snake River
Walla Walla
Astoria
Seaside
Columbia River
Billfish
The Legendary Bigfoot
Kennewick
OREGON MUSEUM of Science and Industry
extinct Plotopterid flightless seabird
WHAT IN THE... SAM HILL'S STONEHENGE MARYHILL
Goldendale
Hermiston
TYLOCEPHALONYX
Columbia River
Portland
Vancouver
Maryhill
Pendleton
Tillamook
Aetiocetus
The Dalles
TO THE JOHN DAY FOSSIL BEDS NATIONAL MONUMENT
Archaeotherium The "Killer" Pig
A FUNKY CHALICOTHERE
Pacific Ocean
A duck
Beaver

N
NE
NW
W
E
SW
SE
S
Kootenay Lake
Montrose
Metaline Falls
IDAHO
Coolin
Priest River
Newport
Loon Lake
Clayton
Deer Park
Blanchard
Spirit Lake
Columbian Mammoth
395
95
195
Palouse
Potlatch
Pullman
270
Moscow
8
Clarkston
Lewiston
Snake River
Hells Canyon National Recreation Area

# WASHINGTON

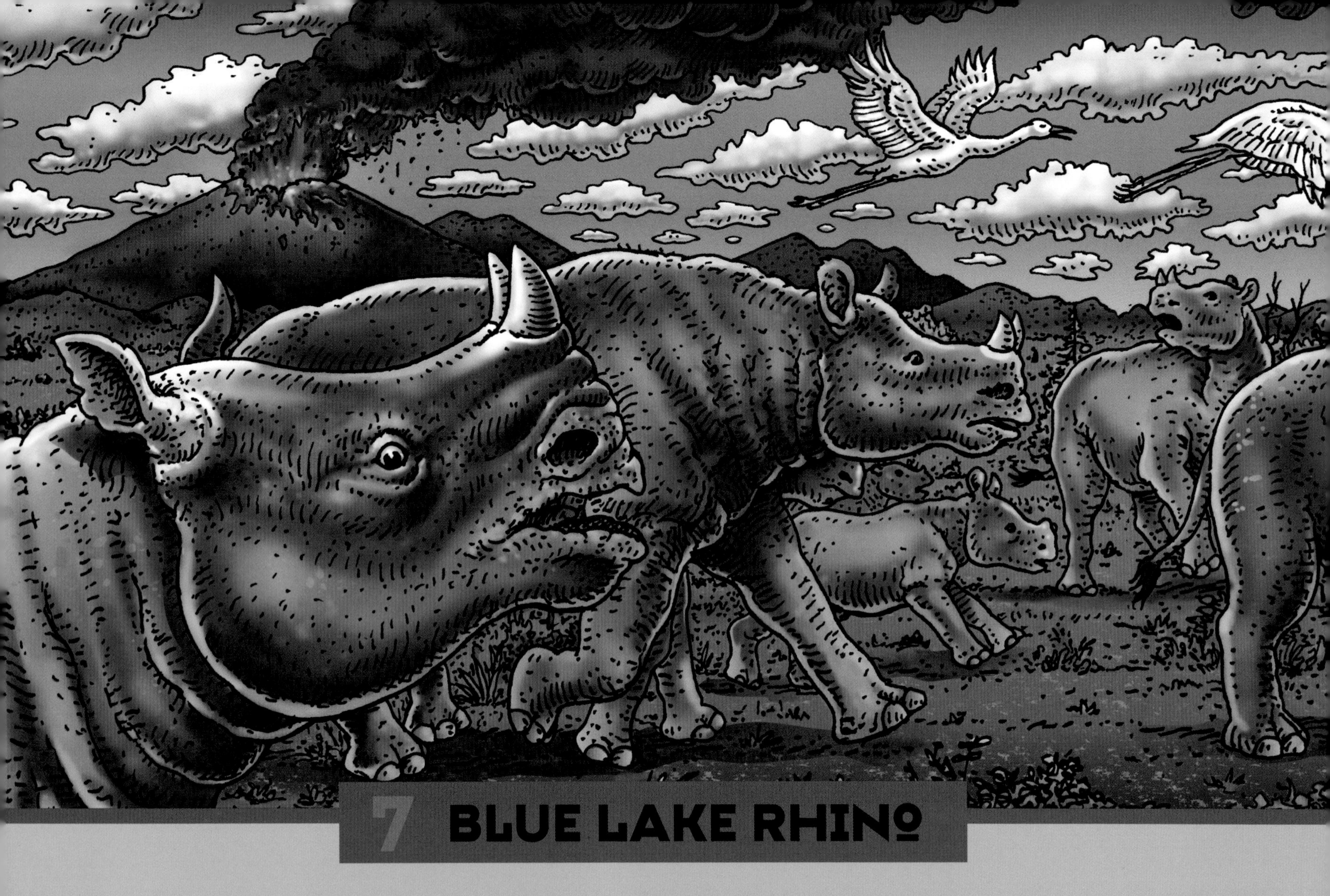

# 7 BLUE LAKE RHINO

In 2009, the Burke Museum in Seattle decided it would do an exhibit based on our *Cruisin' the Fossil Freeway* book. That brought both Ray and me to Seattle several times to meet with the museum staff who were planning the exhibit. At the time, my dad still lived in my childhood home in Beaux Arts Village on the shores of Lake Washington, and using the home as our base, we were in a perfect spot to launch a series of short trips to explore the fossils and rocks of my home state and to revisit some of my favorite spots.

Thirty-two years earlier, I had been a high school junior with a freshly printed driver's license and access to my mom's bright orange Audi. That car allowed my artist friend and mentor, Wes Wehr, and me to move freely about the state in search of fossils and adventures. Wes and I met on June 12, 1974. I was thirteen and he was forty-five. I know the exact date because I recently found a note that my mother had written about the visit. I was a fossil-obsessed teenager, and my mom had heard about an artist

Fifteen million years ago things went from bad to worse for one particular rhinoceros.

Wes did not do what most people did. He did not have a job, he was not married, he did not have a driver's license, he had never driven anything, he did not have a house, and he did not have any visible means of support. What he did have was a persistent curiosity about art, fossils, and interesting people. He was a relentless correspondent who spent hours of each day drinking coffee, smoking cigarettes, and writing letters. His many friends included artists, musicians, poets, actors, writers, philosophers, and scientists. Somehow he had convinced the Burke Museum to take him on as the unpaid curator of paleobotany, an assignment that arose from his love for the beauty of the fossil wood of eastern Washington. His own art was remarkably precise and markedly finite. He used to tell me that he could carry an entire show in his leather satchel. His paintings were tiny austere landscapes, rarely larger than a 3-by-5-inch index card. His drawings, done in India ink with a Montblanc pen, depicted tiny beings that were part kachina and part entomology.

who liked fossils and who was working at the Burke Museum at the University of Washington. At her urging, I called him, and at his invitation, mom and I visited him. Mom's note read, "Interesting day for Kirk. He and I visited an artist (but more importantly for Kirk) paleobiologist Wesley Wehr." The meeting went well. In an odd coincidence, my mom and Wes had been born on the same day in the same year. Both were artists. One had a fossil-obsessed son and one was a fossil-obsessed artist. It was the beginning of a very long friendship.

We became close friends, and I added correspondence and drawing to my tool kit.

Wes introduced me to the magical world of scientific literature. For the last 300 years, science has moved forward and aggregated knowledge through the published papers of scholars. Up until I met Wes, I had been trapped in the popular literature of guidebooks and rock-hounding magazines. He turned me on to the *Journal of Paleontology* and the various publications of the US Geological Survey. With scientific literature in my reach, I was now armed and dangerous. I could learn what the experts knew. Between 1977 and 1989, Wes and I explored Washington, Oregon, and British Columbia, visiting old fossil sites and discovering new ones. We met other fossil enthusiasts, and our finds ended up in the Burke Museum collections. Wes played a huge role in my life, launching both may career in paleontology and my impulse to collaborate with artists.

Wes died in 2004, but his memory is strong. He had instilled a lifelong love of art, science, and road trips in me so it was fitting that, five years after his death, Ray and I were back in Washington State and looking at it with fresh eyes.

We grabbed Burke Museum exhibit designer Andrew Whiteman and set off to inspect eastern Washington in order to gather video footage to support the Burke show and to make the exhibit more about Washington State. It was a glorious sunny June day as we drove east out of Bellevue

Upper: Artist, paleobotanist, and mentor Wesley Wehr.

Lower: "Monster drawings" by Wesley Wehr.

Opposite: A six-year-old Kirk shares his wagonload of rocks with a friend.

toward the Cascades on I-90. Just a few miles into the trip we drove along the southern shore of Lake Sammamish, and just before entering Issaquah we drove past the remains of a roadcut full of childhood memories.

When I was twelve and had learned how to use a library, I began to research fossil sites that were close enough to my home that I could get there and back in less than two hours. That mattered because I could convince my mom to drive me places if it didn't destroy her whole day. Sadly, the guidebooks were pretty blank when it came to local sites, so it came down to trial and error. Then one day, Mom and I stopped at a roadcut on a frontage road above I-90 that was only fifteen minutes from our house. She gave me thirty minutes to check it out.

Most of my previous efforts had come up dry, but much to my amazement and joy, the crumbly cliff was full of fossil clamshells. The rock was soft and wet, and it was easy to pull apart with my bare hands. The clams themselves were so delicate that when I pried them loose, the shells would crumble into fragments, leaving a mud cast of their interiors. The thirty minutes passed in a flash, and pretty soon Mom was honking the horn to let me know that she was done and so was I. We headed home, but I was elated because I now had a fossil place that I could easily access. Mom never minded giving me an hour here and there, and thus was born the one-hour fossil trip. Over the next few years, we went there with some regularity, and each time I learned a little more about the cliff.

I found that if I was very careful, I could get the fossils home unbroken, and, when they dried, they hardened up a bit and became more stable. I discovered that although there were mainly clams, I could find other things if I was persistent. And I was persistent.

One day, I discovered a 2-inch-long shell that looked like a miniature hollow elephant tusk. That night at home when I consulted my books, I was able to identify my first *Dentalium* and learn that is was the shell of a marine animal known as a scaphopod. Not surprisingly, its common name is "tusk shell." This was a curious thing to me because I had never seen one of these shells while beachcombing or in an aquarium. Here was a creature that first presented itself to me as a fossil. This was to become a pattern.

After I had met Wes, I would visit him at the museum where he told me about paleontologists and how they specialized in very particular things. There were people who worked only on fossil wood, or only on fossil leaves. There were even a couple of people who worked only on fossil nuts and seeds. He had figured out who some of these scientists were and had learned that they were happy to receive letters from people who found examples of what they studied.

Mom kept driving me to the roadcut, and my persistence continued to pay off. One day I found a fossil pinecone. Another time I found a fossil nut. This time, Wes and I packaged up the nut and shipped it off to Steve Manchester, the Camp Hancock alum who specialized in fossil nuts. He wrote back and said that it was an extinct relative of the walnut and that he would like to describe it in an upcoming publication on the evolution of the walnut family. I was starting to get the hang of this paleontology thing.

Then one day, I found a bone. I wondered what kind of outcrop would produce clams, nuts, and bones. I gingerly collected the 5-inch fossil and took it home, where I scraped away the soft matrix to expose a bone that looked quite distinctive. I had no idea what it was so I took it to the Burke where Wes introduced me to Bev Witte, a small stooped lady who worked as the museum's preparator. Bev was wonderfully welcoming and immediately identified my bone as one half of a dolphin vertebra. That completely surprised me, and then she asked if I would donate the fossil to the museum because no one had ever found a dolphin bone in that place before. She made the point that my find would be of more use to science if it was in a museum than if it remained in my bedroom. It was my first find of any significance and the start of my life of collecting for museums.

As Ray, Andrew, and I drove past the outcrop, I pointed it out and told the story of the tusk shell, the dolphin, and the walnut. It was amazing to think that thirty-five years had elapsed since I had made those finds. Now I was a geologist and I knew a lot more about the outcrop. I knew that it was the Blakely Formation, a marine unit of Oligocene age, and that my fossils were about 30 million years old. The other thing I knew was

## The Seattle Fault

Seattle has earthquakes, and I can even remember feeling one in 1965 when I was a kid. But the size and danger of Seattle's earthquakes has only come to be known in the last thirty years. It turns out that Seattle has a variety of different kinds of earthquakes, and some of them are pretty significant. The Seattle Fault was first recognized as a potential problem in 1992 when scientists literally connected the dots and showed that the Seattle Fault stretches from Bainbridge Island to Alki Point (the spot where Seattle was founded in 1851) to Mercer Island and across Lake Sammamish. The scientists showed that the fault had ruptured 1,100 years ago in a huge earthquake that uplifted several areas, caused tsunamis in Puget Sound, and sent huge patches of forest sliding into Lake Washington (you can still find these submerged standing forests on the bottom of the lake). In the time that elapsed between my childhood fossil forays, the geology of Washington has become impressively dangerous. Actually, the geology hasn't changed at all. What has changed is how much we know about it.

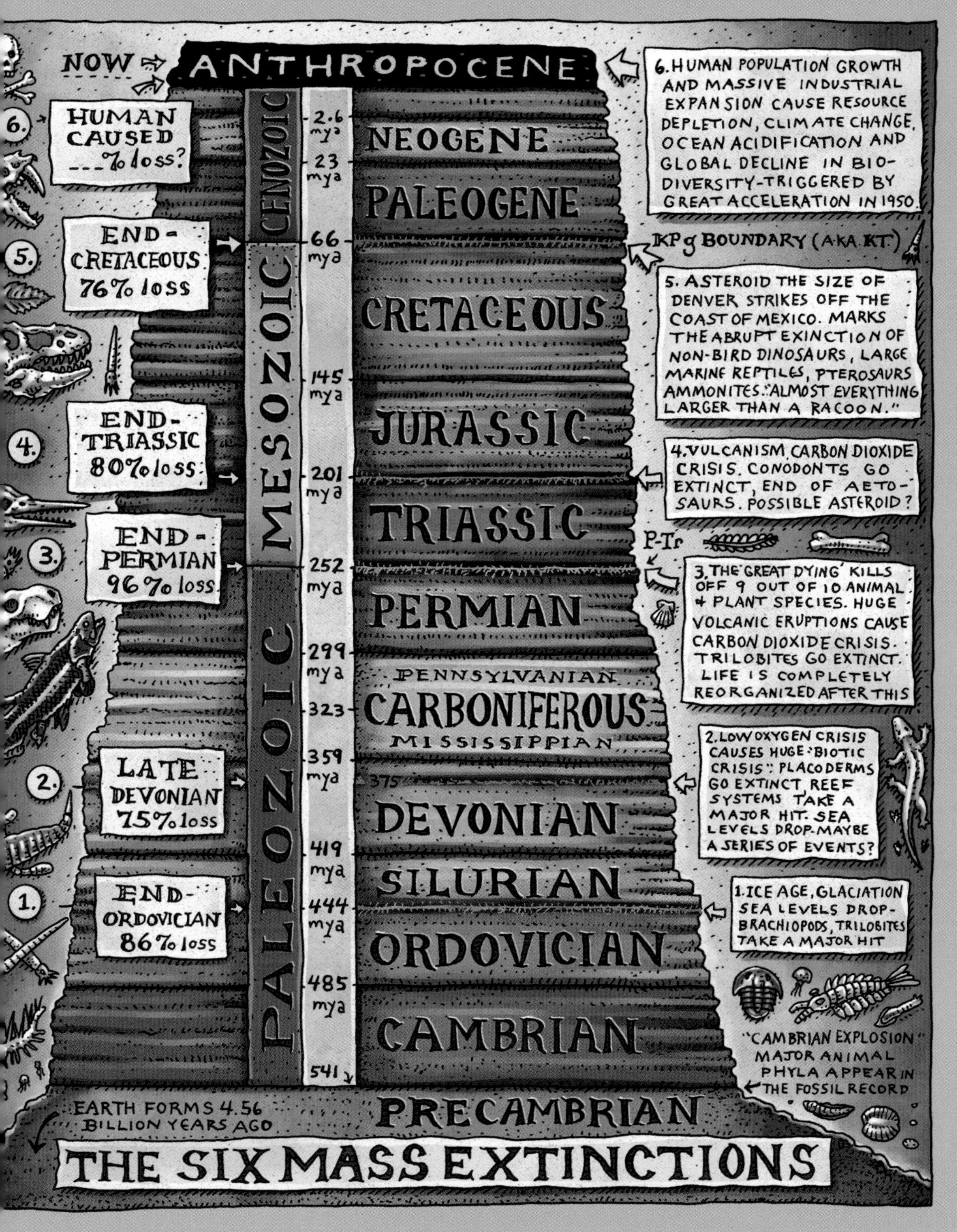


that the outcrop lay just to the south of an east–west feature known as the Seattle Fault.

We continued up into the foothills of the Cascades and finally crested them at Snoqualmie Pass. At a mere 3,015 feet above sea level, this pass is home to Seattle's most accessible ski area, but its low elevation means that you are often skiing in the rain. It was this fact that cured me of skiing at an early age.

We stopped at a pancake house for breakfast, and because Andrew wanted to make a small film about pancakes. For years, I have searched for good metaphors to describe geology to people who don't know or care about it. Again and again, I find that food metaphors work the best. Andrew wanted something for the exhibit that would help visitors understand how sediments became rocks, and corpses became fossils. Pancakes make good analogies for geologic formations both because they are flat and have a certain thickness and a certain lateral dimension and because stacks of pancakes make a good analogy for stacked layers of rocks that form the geologic record.

Think for a moment about pouring pancake batter on the flat griddle. As the liquid batter heats and cooks, it stops flowing and solidifies into a pancake. It turns out that eastern Washington is underlain by a whole lot of rock that started its life something like pancake batter. The entire Columbia Plateau of the southern half of eastern Washington is made of stacks of layers of volcanic rock known as basalt that erupted in a series of lava flows between 17 and 14 million years ago. These lava fields extend into eastern Oregon and westernmost Idaho, and in places their total thickness is nearly 6,000 feet. Individual lava layers range in thickness from a few tens of feet to a few hundred feet, and this means that the original lava flows were this depth as they flowed out of cracks and along the surface, sometimes traveling at speeds of up to 20 miles an hour. The temperature of liquid lava is around 1,000°F, which means that huge, incandescent rivers of molten rock were a common and deadly sight in eastern Washington. This could not have been a pleasant thing for the plants and animals that had the bad luck to be in the path of the molten rock rivers.

The view down Interstate 90 of the Columbia River and its canyon cut into Miocene basalt layers.

As we rolled off the eastern slopes of the Cascades, we began to see roadcuts composed of the black volcanic rock known as basalt. This rock often forms distinctive vertical five-sided pillars that look like giant black crystals. The pillars (or columns as they are more commonly called) formed as the liquid lava started to cool. As it cooled, the volume of the lava decreased, creating stress in the hardening rock that resulted in cracks. The intersecting cracks form the pillars.

A few miles past Ellensburg, we started the long, gentle decline down to the Columbia River. The mighty Columbia is the largest and longest river that flows into the Pacific Ocean from North

America. Its headwaters are in British Columbia more than 1,200 miles from the ocean, and it winds a very long path before entering the northeastern corner of Washington and flowing across the state, then along the Washington–Oregon border, and finally out to the Pacific at Astoria.

The Columbia was once home to the largest salmon run in the world, but that ended with the exploitation of the river for hydroelectric power that began in the 1930s. I can still remember my first grade teacher regaling us with tales of how he worked on the construction of the Grand Coulee Dam. Now, fourteen different dams block the river, creating a series of long, flat lakes instead of a rushing wild waterway. The dams generate incredible amounts of electricity, but they block the salmon from spawning in the upper thousand miles of the river.

A petrified log entombed in basalt was once a log buried in molten lava.

By the 1940s, the banks of the Columbia near Hanford became the site of a secret nuclear facility that eventually produced the plutonium used in the atomic bombs dropped on Hiroshima and Nagasaki. When I was sixteen, I was selected as part of a group of high school students to tour the Hanford site and learn about the future safe uses of nuclear power. I remember that they gave me a marble that had been radiated as evidence that nuclear power was okay and that we would soon be seeing more nuclear power plants than hydroelectric dams in Washington. A few years later, that idea went away in a spectacular bankruptcy, and we were back to hydroelectricity.

But it wasn't salmon, dams, electricity, or atomic bombs that drew Ray, Andrew, and me to the valley of the Columbia. As we drove the last few miles down the long hill toward the river, I pointed out tubular holes in the basalt outcrops as we whizzed past. Ray wasn't really sure what I was pointing at and missed the first few. Finally, I pulled over and stopped next to one of the bluffs. There along the base of the cliff we could clearly see several of the curious tubular tunnels. We got out and walked over to the cliff and peered into one of them. About 3 feet back from the face of the cliff was the clear cross section of a perfectly petrified log. A petrified log in a basalt cliff could mean only one thing: a long time ago, a tree had been entrained in a lava flow.

The fossil logs in the Columbia basalts were first noticed by highway workers in 1927. In 1931, George Beck, a music professor at the Central

Washington College of Education, learned about the fossil wood and began to study it. The logs were so well preserved that every detail of the cells were visible, and it was possible to slice thin sections of the logs and identify the type of tree they came from. Beck involved students in the effort, and they began to scour the hills in search of logs weathering out. He was amazed to find that the logs represented many different species of conifers and broadleaf trees, none of which live on the barren slopes today. In 1932, Beck discovered a petrified log that had the distinctive cellular structure of *Ginkgo*, the maidenhair tree. *Ginkgo* is a tree that today lives only in China, Japan, and Korea, where it is widely planted near monasteries and temples. Beck realized that the site was significant and approached the state government and petitioned for the site to be protected. In 1935, the Ginkgo Petrified Forest State Park was established.

Wes Wehr had started to correspond with Beck in the 1950s and eventually started coming east to search for fossil wood himself. Beck told him stories about the local competition to find petrified logs. The logs were embedded in the basalt, but as the ground weathered, distinctive fossil wood chips at the surface could be an indication of a log in the weathered basalt below. Basalt is a hard rock and digging for logs in the basalt was tremendously tough work.

Beck found himself in competition with a man named Frank Bobo who was in the habit of sprinkling chips of petrified wood in random places in hopes of causing competitors to dig in the wrong places and not rustle his logs. The prize for all this hard work was some of the most beautiful petrified wood in the world. We had entered the weird world of the paleo-lumberjack.

We exited the interstate in the tiny town of Vantage, which is located on the west bank of the Columbia above the Wanapum Dam. What had been a beautiful canyon was now a large lake that had formed in 1959 after the dam was built. The state park is a single building with a few exhibits about the story of the fossil wood and some nearby petroglyphs. Near the entrance to the park there was a big sign for the Ginkgo Gem Shop.

This Washington state park is named after a Miocene fossil tree.

I had warned Bill that we would be stopping by, and he and his wife and two kids were waiting when we pulled into the lot. Bill had bought out an Arizona rock shop, so the lot was filled with giant concrete dinosaurs and huge red petrified logs from the Triassic rocks on private land near Petrified Forest National Monument by Flagstaff, Arizona. Apparently, Bill was selling so much wood that he

Above: Bill Rose shows off his massive rock saw.

Right: A petrified log.

Bill Rose and his wife, Dee, have operated the shop for more than thirty years, and Bill is a man completely focused on petrified wood. I've made a point of stopping by his shop for the last three decades, and he hasn't seemed to age at all. The Ginkgo Gem Shop is a classic example of the western rock shop. Back in the 1960s, just about all western towns had rock shops, but they are now an endangered species. Bill's joint is unique because of his focus on the local product: Miocene fossil logs. He mines them out of the loose basalt rubble, then slices off round slabs with diamond-bladed rock saws. He then uses finer and finer polishing grits and powders to bring the face of the slab to a high luster. The result is an extraordinary view into the lost forests of the Columbia Plateau.

needed to start importing it from other states. The Arizona logs were nearly 3 feet in diameter, and Bill had used them as an excuse to buy the biggest rock saw I had ever seen. The circular blade was 6 feet in diameter, and with it, Bill could saw through any petrified log he had.

He took us into a series of workshops and storerooms, each full of old rock saws and cut slabs of wood. Every space was coated in rock dust; the whole place looked like a volcano had erupted and belched volcanic ash all over his inventory. Time after time, he would wipe away the dust to expose a glassy polished slice of fossil wood. He knew his fossil wood well and announced the species of tree with every slab: honey locust, Douglas fir, oak, maple, hickory, elm, sour gum, sequoia, ash, pine, cherry, apple, sweetgum, and yes, *Ginkgo*. The forest that had once grown here was a diverse temperate broadleaf forest with a few conifers thrown in for good measure. More than forty species have been found here. Like the forests of eastern Oregon of the same age, this forest was more similar to the deciduous forests of New England and northern China than it was to what grows in Washington today.

While these forests left their logs in the lava, there are sites between the lava layers that also preserve their leaves. One such site at a place called Clarkia in western Idaho exhibits an extraordinary feature. The wet, dense clay at Clarkia has excluded oxygen since the Miocene, and when you split open a block of Clarkia clay, it is actually possible to see a green leaf. Within moments though, the leaf reacts with the modern atmosphere and turns black.

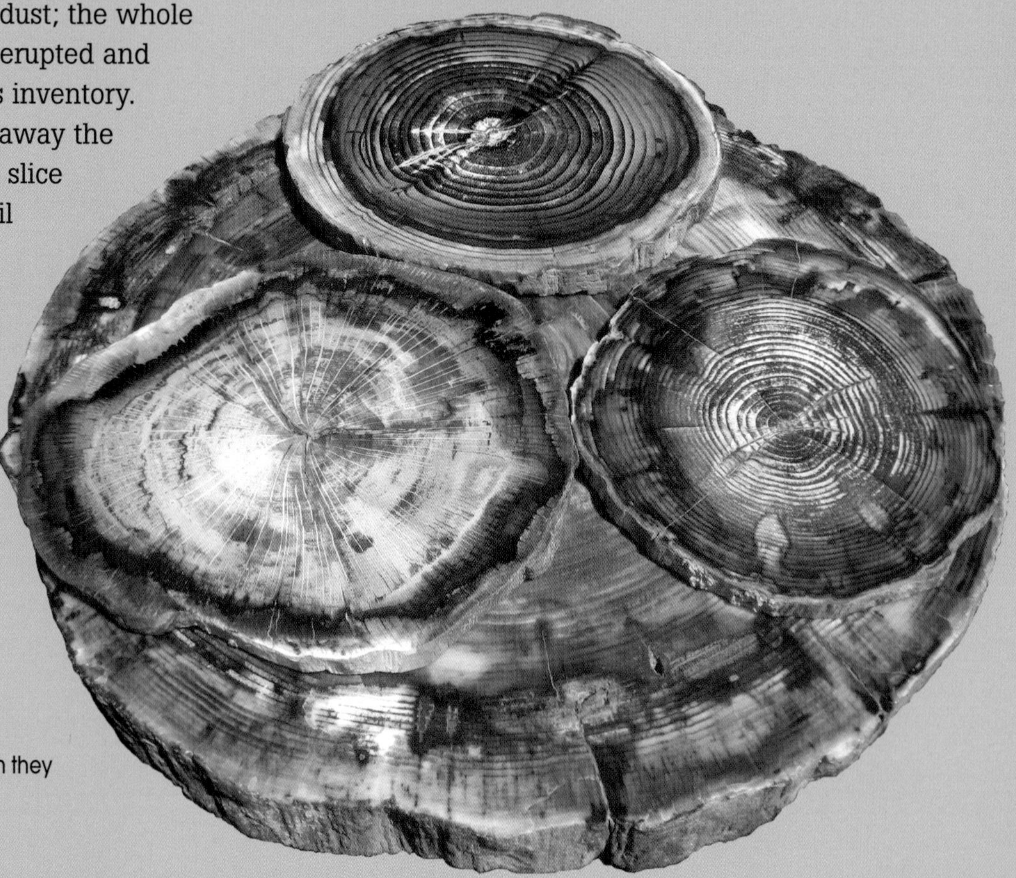

Above: Bill Rose with a perfect slab cut from an agatized Miocene log.

Right: The tree rings are so well preserved that we know how old the trees were when they were buried in lava 15 million years ago.

We bought a few slices of fossil wood and headed across the river. The road climbed steeply out of the canyon and we drove past cliff after cliff of basalt. Ray's eyes were now trained on the roadside, looking for the round holes that signified fossil logs. He had no luck, and we crested the hill near the town of George and headed north to Ephrata and Soap Lake.

At Soap Lake, we dropped back into a big canyon and headed north toward the Grand Coulee Dam. Unlike the canyon at Vantage that had a huge river in it, the canyon between Soap Lake and Grand Coulee had no river. Instead, the floor of the canyon had a series of small lakes. To a geologist's eye, a canyon without a river is an odd thing indeed, since it is rivers that make canyons.

I was by no means the first geologist to take note of this odd situation. In 1905, a young man named J. Harlen Bretz graduated from college and moved to Seattle to become a biology teacher. The geologic grandeur of the area gripped him, and his interests quickly turned to geology. The concept that the northern half of North America had recently been covered by a 2-mile sheet of ice was an idea that had taken root on the East Coast in the 1870s, and it was pretty clear that Puget

Blue Lake.

Sound had been formed by the southern extension of this massive ice sheet. Bretz decided to go back to school and become a geologist, and he enrolled in the University of Chicago and continued to study Puget Sound. He became a professor at the University of Washington in 1913 and published a detailed report entitled *Glaciation of the Puget Sound Region*. In it he demonstrated an acute eye for landscape and an intense love of fieldwork. He was a tough critic and bluntly described the work of one of his predecessors as "based on too little fieldwork, and much of it is in error."

In 1922, Bretz found himself in Soap Lake trying to understand the enigma of the canyon that had no river. North of Soap Lake the canyon broadens, and a huge cliff cuts across the middle of it, creating a feature called the Dry Falls. To the north of that stretches the Grand Coulee, a long canyon that intersected the Columbia River near the site of the Grand Coulee Dam. Elsewhere to the east, the surface of the lava flows are disrupted by huge, erosive divots. Initially, it seemed like the Grand Coulee might just have been an abandoned canyon of the Columbia, and the Dry Falls once Washington State's version of Niagara. But Bretz looked past the obvious and saw the extraordinary. Armed with his understanding of glacial geology and his willingness to undertake extensive fieldwork, he came to the conclusion that eastern Washington had been inundated by huge floods of glacial meltwater that had eroded the landscape with unprecedented fury. He published his idea of the "Spokane Floods" in 1925 and was immediately attacked by colleagues who would not see what Bretz had seen. It took him forty years to win his argument, but win it he did.

The event that triggered the floods was the advance of a lobe of ice from Canada down to what is now Lake Pend Oreille in northern Idaho. The lake bed itself and the surrounding cliffs were gouged out by earlier glaciers. As the south-flowing glacier filled the lake, it blocked a west-flowing river (now known as the Clarks Fork) and created a lake that backed up more than 100 miles and all the way to the present site of Missoula, Montana. Eventually, the lake grew to be

Above: Ray rows across Blue Lake.

Right: Ray ascends the talus slope.

more than 2,000 feet deep and contained more than 600 cubic miles of water. The depth and pressure of the water eventually floated, and then shattered, the ice dam, releasing a wall of water that was nearly one-half mile high.

This massive flood roared across eastern Washington at speeds of up to 45 miles an hour and literally resurfaced the state as it went. Eventually, the flood washed into the valley of the Columbia and roared out to Astoria before carving a now submerged canyon at the edge of the continental shelf and spending itself in the Pacific Ocean.

With the pressure released, the glacier pushed back into the lake and created a new dam. The river built up behind the new dam, and after about fifty years, another massive dam rupture created yet another massive flood and started the process all over again. It is now widely accepted that this extraordinary event happened at least forty times between 15,000 and 13,000 years ago.

Most people don't think of vast flows of molten lava, dense temperate forests, or gargantuan floods as they drive across the broad barren expanses of eastern Washington, but that is this place's reality.

I had one more trick up my sleeve as we drove north from Soap Lake. I had recently gotten a tip about how to find one of the oddest fossils in North America. We drove past Soap Lake, Little Soap Lake, Lenore Lake, and Alkali Lake and finally pulled over at the north end of Blue Lake and followed the signs into Laurent's Sun Village Resort where Ray and I plunked down $9 to rent a rowboat. Andrew had his own kayak, and my friend Cathy Lou Brown had driven down to meet us. It was a hot afternoon and the lake looked more green than blue as we set out in our two boats to cross it. We were going to make a serious effort to find the legendary Blue Lake Rhino.

This story started in 1935 when two couples, the Frieles and the Peabodys, from Seattle were poking around the lava cliffs above Blue Lake in search of petrified wood.

They were likely using the same search image that had worked for us in roadcuts: look for round holes in the cliffs. On this particular day, they found a particularly large hole. Mr. Friele crawled into the hole, expecting to find fossil wood. Instead, he found several fragments of fossil bone and part of a fossil jaw with parts of six teeth. Mrs. Peabody took these bones to the University of Washington, who sent them to George Beck, who visited the site later that year. Beck found some more bones and shipped them to Chester Stock at Cal Tech, who identified the jaw as belonging to a Miocene rhinoceros called *Diceratherium*. In 1948, a crew from the University of California–Berkeley Department of Paleontology visited the site and made a plaster mold of the inside of the cavity. When assembled, the mold had the very distinctive shape of a large and somewhat bloated four-legged rhinoceros lying on its back. The rock that formed the walls of the cavity is known as pillow basalt, a type of rock that forms when lava erupts or flows into water. The obvious conclusion was that a rhino (and we'll never know if it was alive or dead at the critical moment) was in a shallow pool or stream when it was overrun, entombed, and no doubt baked in flowing lava.

Eventually, the lava cooled to stone and was buried by many other similar layers. Then 15 million years passed, and the Spokane Floods eroded canyons and gullies and miraculously eroded open a hole at the tail end of the beast. Then 13,000 years elapsed and the Frieles and the Peabodys arrived at the cliff and found the hole. Now seventy-four years later, we had arrived at the base of the cliff and were surveying it in an attempt to find a hole on the cliff face. No such hole was apparent, but we could see an inviting ledge about 200 feet up the cliff. Someone had painted a white letter "R" just above the ledge, and we took this as a very good sign that we were on the right track. We began to scramble up the steep talus slope composed of jagged basalt boulders. It was hot, but we were in the shade and the day was calm. I was really excited.

At the top of the talus slope we encountered the base of the cliff and were confronted with a little zone of treacherous verticality. Andrew and Cathy worked their way up the cliff and disappeared above us. I followed and was making my way up when I heard the smallest of whimpers. I glanced below me to see Ray completely prostrate on the cliff and utterly unable to move. The Troll had frozen.

I worked my way back down to where he was to assess the situation. It was not good. Ray was unable to go up or to go down. This presented a problem for our little expedition. I had not been aware that Ray had a fear of heights and neither, apparently, had he until it struck him. I maneuvered myself into a position immediately below him and used my rock hammer to dig a flat bench where he could find a stable footing. Now that I had stabilized the Troll, I began to soothe him with the thought of the discovery that awaited us above. After fifteen minutes, his equilibrium returned and we gingerly made our way up to a ledge that was the width of a narrow sidewalk and ran for about a hundred yards along the cliff face. The drop-off below the ledge was not insignificant, and we all walked cautiously as we searched for the legendary hole.

We found several small holes that must have once contained petrified logs, but the large rhino hole was nowhere to be found. Then we found the letter "R," but there were no holes

FEARING THE WORST
RHINO
TROLL BRAVELY STRUGGLED UPWARDS

Cathy Lou Brown peers out through an opening that was once the back end of a rhinoceros.

near it. We had come so far and we were so close, but we were truly stumped. After about fifteen minutes of searching we were about to give up, when Cathy noticed a plastic ammo box crammed in a small cavity at the base of the back wall of the ledge. She pulled it out and opened it up. It was a geocache with a series of notes. Several notes celebrated the success of their authors in finding the rhino. Several others expressed exasperation and frustration at having come so far and having not found the rhino. Then we found a note that said:

*COOL TREASURE BUT CAN'T FIND THE RHINO ... KNOW IT'S CLOSE BUT. RICK VOGEL BELLINGHAM WA NOW IN GREENVILLE SC. P.S. WOULD BE NICE IF SOMEONE WOULD DRAW MAP TO THE RHINO'S LOCATION. FOUND IT! STRAIGHT ABOVE THIS CACHE. COOL.*

We looked straight up about 9 feet above our heads and there was the hole. We were elated, and I was just a bit terrified. As much as I wanted to crawl into the rhino, a 9-foot climb above a narrow ledge above a very long drop did not appeal to me. This was not even an option for Ray. I paused, and in that moment Cathy scrambled up the cliff and disappeared into the hole. A moment later, she reappeared, her impish face grinning from the rhino's butt.

Cathy luxuriated in the rhino for about fifteen minutes, allowing me time to screw on my courage. I had not come this far not to crawl into the rhino's rump. So up I went and in I went. Mission accomplished. For a man who loves fossils like a kid, this was a real moment for me. Ray chose not to test his vertigo, instead staying on the ledge below and beginning to compose a song about the sad fate of the baked rhino. With our mission accomplished, we worked our way down the cliff and rowed back across the lake. Nine dollars well spent.

The next day we crossed the Columbia at the Grand Coulee Dam and entered the remote part of Washington known as the Okanogan. We entered the Colville Indian Reservation and drove north along the forested floor of the deep and exquisite San Poil River canyon. We drove north for an hour and entered the little gold-mining town of Republic. It was in this town in the summer of 1977 that I kicked a rock that changed my life.

I got my driver's license in the fall of 1976, and my friend Wes and I immediately began planning a road trip for the following summer. My mom was delighted to be relieved of the task of driving us and happily gave me the keys to her car. The idea was a big clockwise loop that would take us north from Seattle to the Skagit Valley, east to the Okanogan highlands, and south to the John Day Country of eastern Oregon before stopping in Portland on the way back to Seattle. Wes had been reading the research papers of a USGS paleobotanist by the name of Roland Brown who had passed through eastern Washington in the 1920s. Our goal was to visit some of his fossil sites and see if we could find new ones. It was my first time as the responsible party on a long road trip.

I drove and Wes smoked. He was a slight and very quiet man. He spoke a lot, but he spoke in a whisper. To hear him I had to lean across the stick shift. We departed Seattle and crossed the North Cascades Highway, an amazing mountain road that had only been completed five years earlier. In the town of Winthrop, we found a roadcut

THINGS WENT FROM BAD TO WORSE FOR THE RHINO ON THAT FATEFUL DAY OH SO LONG AGO.

that had 120-million-year-old Cretaceous fossil ferns, conifers, and broadleaves. Twenty years later, Ian Miller, a Winthrop kid, would complete a PhD at Yale University and demonstrate that these fossils from the Winthrop Sandstone had grown at the latitude of Mexico, and their fossils had been shifted north on a block of exotic terrane only to be emplaced in northern Washington as stunning evidence for what has come to be known as the Baja–BC hypothesis.

Wes and I continued east through Twisp and Omak and into the Okanogan, arriving in Republic just after noon. Republic was a quiet town located a few miles south of an operating underground gold mine. There was an obvious roadcut at the south end of town, and we spent a few hours turning over rocks and finding nothing. We stopped at the library and asked if anyone knew about fossils. No one did. Wes had received a letter from a Republic woman saying that she had found a perfect fossil flower, but that seemed preposterous, and, in any case, we could not find her.

We stopped at the café and ordered milkshakes and pondered our next move. Since Republic seemed like a bust, we figured that the best bet would be to head straight south into Oregon and see if we could get to the John Day Fossil Beds. My orange Audi was parked on the side of Main Street at the south edge of town. Wes wanted to smoke a final cigarette before we left. I walked around the car to check the tires and noticed that the dirt at the side of the road had pieces of very fine-grained tan shale. I kicked one. It was about 2 inches by 2 inches and it popped open like a little book. And there on both facing surfaces was the most exquisite fossil conifer sprig. In truth, it looked like one of Wes's little drawings. It was an

Right: A view of the town of Republic from the Boot Hill fossil site.

Opposite left: Paper shale.

Opposite right: A perfect fossil of the flower *Florissantia quilchenensis.*

amazing fossil and it had been lying in the dirt 6 inches from the asphalt of the road.

I looked at Wes, he looked at me, and we wordlessly began to unpack our tools. Within minutes, it was quite clear that the drainage

ditch was full of treasures. The small flat pieces of shale were sturdy, but they split easily and they were full of gorgeous fossils. This kind of rock is called paper shale, and opening the slabs was like turning the pages of a book. We quickly began to accumulate conifer shoots, alder cones, pine needles, alder leaves, and insects, and then – a perfect five-parted flower. The woman had been right.

We dug in the ditch until the sun went down and then camped in the city park. In the morning we were back at it, this time on the other side of the road where some actual bedrock was exposed and we could pull out larger slabs. We could not believe our luck. We had stumbled on an amazing fossil site in the most unlikely of places. We stayed for another day and then headed south to Oregon. The rest of the trip was great, but the discovery at Republic changed the future for me, Wes, and the town.

When we got back to Seattle and unpacked our treasures, Wes found that he could easily photocopy them (they were on paper shale after all), and we sent a stack of copies to Jack Wolfe, a paleobotanist (and former Tom Bones protégé) who lived in Menlo Park, California, and was Roland Brown's successor at the US Geological Survey. Jack wrote back quickly to tell us that our fossils were about 50 million years old and represented the high-elevation vegetation of the Eocene epoch. They were rare, important, and of extremely high quality.

Over the next several years, Wes and I returned to Republic many times and mined the corner lot, filling cabinets at the Burke Museum and shipping boxes of fossils to Menlo Park. As the Republic collection grew, so did the diversity of its fossil flora. In 1987, Jack Wolfe published a *USGS Bulletin* about the site, naming several new species of fossil plant and establishing the Republic site in the scientific literature. In 1989, the town of Republic opened a small museum named the Stonerose Interpretive Center and began offering people the opportunity to dig fossils with the stipulation that any of significance would be retained by the museum for scientific study. Tourists began to visit Republic to see its fossils, and by the mid-1990s, the little museum was seeing more than 10,000 visitors a year. Wes became the resident expert and liaison between the town of Republic and the Burke Museum. Other scientists heard about the site and came to see its gorgeous fossils.

As Ray, Andrew, and I drove into town, the Stonerose Interpretive Center was getting ready to celebrate its twentieth anniversary. Cathy Brown, who had joined us at the Blue Lake Rhino, was the manager of the Stonerose Center, and she had organized a barbecue for our arrival. Stonerose had discovered a second and larger fossil deposit just on the north side of town, and dozens of kids were swarming the slope with screwdrivers, hammers, and juice boxes. There were whoops of joy as real kids found real fossils in a real town. It felt really good to be home.

That night Ray and I inspected the collections of fossils that the museum had high-graded from the kids. Here was a world-class collection of 50-million-year-old fossil plants, insects, and fish that represented an ancient high-elevation world. One of the fish from this site is *Eosalmo driftwoodensis*, the earliest known member of the salmon family. Republic is only 40 miles west of Kettle Falls, a spot along the Columbia River where Native Americans used to net

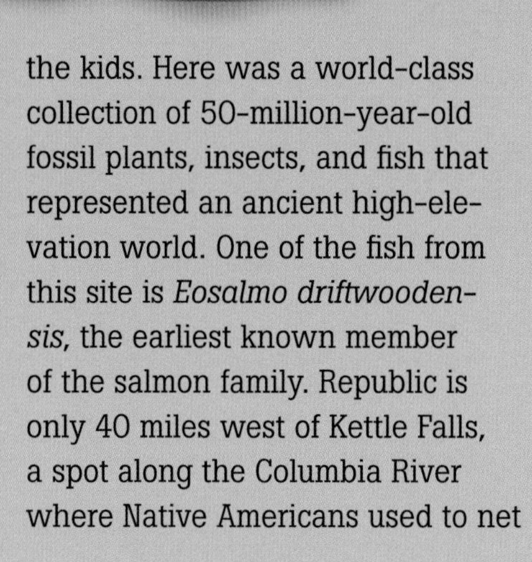

Upper: The dawn redwood, *Metasequoia occidentalis*.

Middle: Fossil leaves of elm (*Ulmus*) and the Japanese scholar tree (*Cercidiphyllum*).

Lower: A fossil pinecone (*Pinus*).

Chinook salmon that were headed north into British Columbia to spawn. Those falls are now submerged under the lake formed by the Grand Coulee Dam, but both Republic and Kettle Falls are places that demonstrate the long history of salmon in the Pacific Northwest.

The full suite of plant species from Republic now numbers well over 200, making it one of the most diverse fossil floras known in the world. There were incredibly beautiful fossils and even some slabs with as many as eight fossil flowers – a literal fossil bouquet. The site contains the remains of a number of plants that still live in the northwest like alder, fir, cedar, birch, and pine. It also includes fossils of the rose family, which includes cherry and apple trees as well as blackberries and currants. Today, Washington State produces nearly half of the world's apple crop and more than half of its cherries. It did not take Wes long to make the connection between the fruit orchards of eastern Washington and the ancient orchards of Republic.

The fossil plants of Republic also carried a surprising twist. Many of them are extinct in North America but have living relatives in China, Japan, and Korea. These included not only *Ginkgo*, or the maidenhair tree; but also *Cercidiphyllum*, the Japanese scholar tree; *Koelreuteria*, the golden rain tree; *Pseudolarix*, the golden larch; and many others. The appearance of Asian trees as North American fossils is strong evidence that there used to be a continuous forest from Washington to Alaska and across into Asia. Our trip was carrying us north, and we would see much more evidence for this as we carried on.

The Stonerose fossil site is open for digging, so we checked out hammers, gloves, and work glasses from the museum and joined the horde of happy kids who were banging away on the hillside. Fossils are common enough that no one went away without having had the satisfaction of finding an Eocene treasure, and even Ray and Andrew found fossil flowers.

Our trip to eastern Washington had been brief and we left a lot of fossil sites unvisited, but the western half of the state beckoned, and the next day we headed back to Seattle and Puget Sound.

THE FLOWER COVERED LAKE HIGH ABOVE THE LAVA

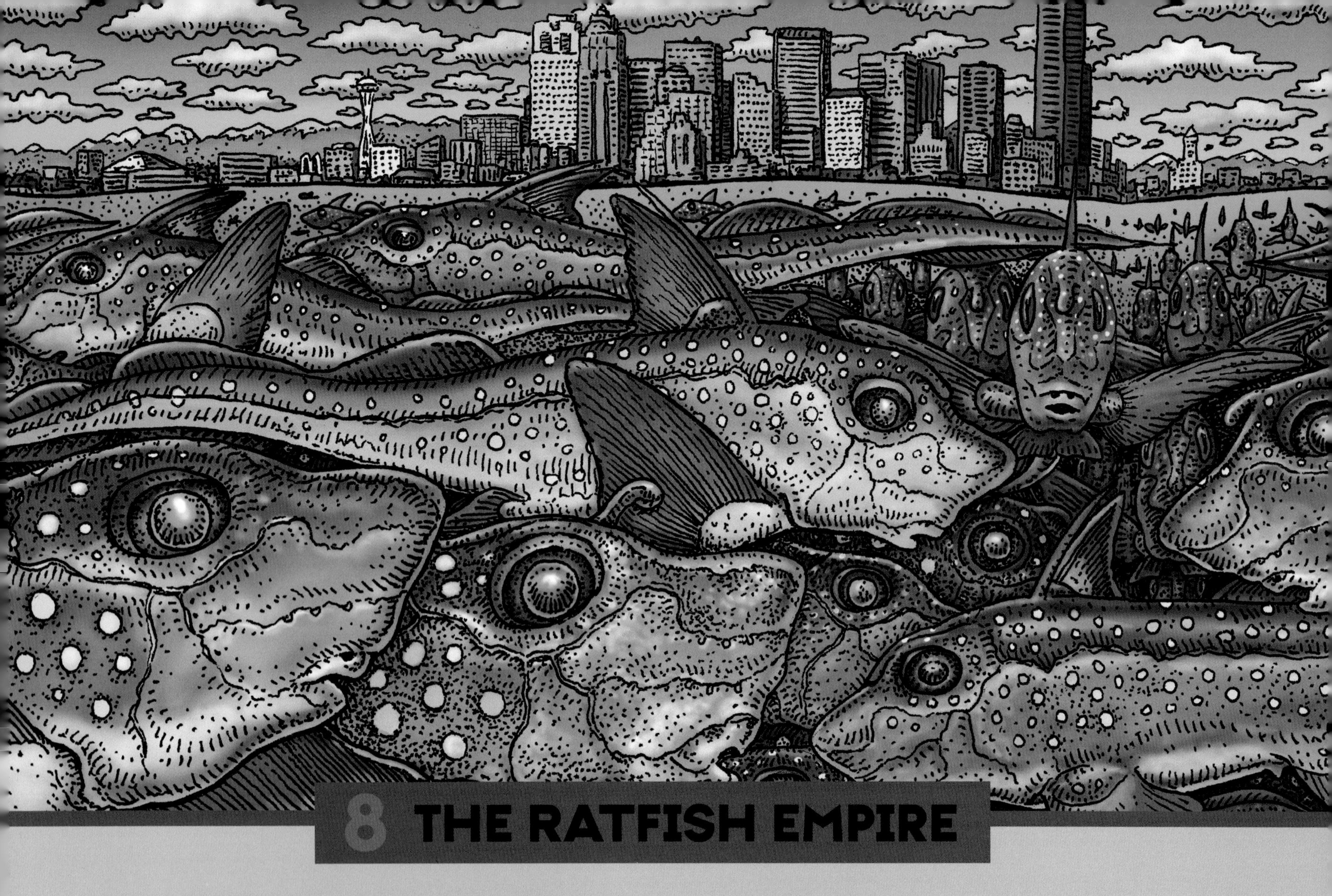

# 8 THE RATFISH EMPIRE

When viewed from far above, the Emerald city of Seattle is surrounded by long bodies of water. Some are filled with saltwater like the many channels of Puget Sound and others are freshwater lakes like Sammamish and Washington. It only occurred recently to me to wonder why such a landscape could exist. The answer to that question is complicated, and it has to do with a surprising history.

My family moved to Seattle in 1961 when I was just nine months old. At the time, it was the rainy, little-visited corner of the nation. My parents loved it because of the mountains and the sailing. The mountains of Seattle are remarkable, especially when the skies clear and you can see them. To the west is the jagged range of the Olympics, a foreboding massif often clad in thick clouds. To the east are the Cascades, pimpled with volcanoes: Mount Baker to the north and the impossibly large Mount Rainer to the south. Rising essentially from sea level to 14,410 feet and festooned with twenty-six glaciers, Rainer is probably the biggest single thing in the lower forty-eight states. It is an active volcano that

The spotted ratfish, *Hydrolagus colliei*, has come to be the most common fish in Puget Sound. They are members of the ancient Chimaeridae family and are related to sharks.

erupted several times in the 1800s and most recently in 1894. The menace of this mountain monster was invigorated when its lesser cousin, Mount St. Helens, exploded in 1980.

Explored first by the Greek Juan de Fuca in 1592, then by the Brit George Vancouver in 1791, and finally in 1841 by the American Charles Wilkes, the Puget Sound country was already home to clans of the Klallam, Skokomish, Squaxin, Puyallup, Nisqually, Muckleshoot, Duwamish, Tulalip, and Stillaguamish people who lived in cedar houses and feasted on the salmon that spawned in the myriad rivers, streams, creeks, and rivulets that emptied into it.

In 1851, a small group of white settlers made their way into Puget Sound by boat and landed at Alki Point in what is now West Seattle. Within a year they had moved into Elliott Bay and started to build what would become Seattle. They found the fishing excellent, and there were endless oysters for the taking. The real prize was the incredible lowland temperate rain forest composed of giant western red cedar, enormous Douglas fir, western hemlock, and Sitka spruce. These forests were some of the most amazing to ever grace the planet. Growing in fertile soil with ample rainfall, some of the Douglas fir were more than 300 feet tall. Soon schooners were making the run to San Francisco carrying barrels of oysters and rough cut timber.

In 1897, the discovery of gold in the Yukon funneled thousands of would-be miners through

North America's Last Glacial Maximum

Seattle as they headed north to Skagway, Alaska. This event did for Seattle what the 1849 California Gold Rush did for San Francisco, changing it from a town to a city. In 1909, Seattle celebrated its growth by hosting the Alaska-Pacific Exposition at a site on Lake Washington that would later become the University of Washington. A few years later, Harlen Bretz arrived and got busy answering my question of why Puget Sound is configured the way it is.

With the same intensity that he would later apply to Channeled Scablands of eastern Washington, Bretz walked the shores, valleys, and mountains surrounding the Puget lowlands in search of clues. He was well aware that big ice sheets had been involved, and his careful work showed just how big and extensive those ice sheets had been. By mapping the margins of the glacial deposits, he could tell how high up the mountains they rode, and in short order he could demonstrate that Seattle itself had relatively recently been underneath 3,000 feet of ice. The steep and rounded hills of Seattle and the linear lakes and saltwater channels were all sculpted artifacts of the massive ice.

This discovery meant that the ancient forests of Puget Sound were less than 10,000 years old. Many of the trees lived for more than 1,000 years, so in a relative sense, these forests had not had a long history. And unfortunately for them, they were easy money, so logging surged. Between 1905 and 1938, Washington State was the leading timber producer in the nation. In 1947, the invention of the chainsaw changed the game even more and the old-growth forests came under intense pressure. By the mid-1960s, most of the old-growth forest in the Puget Sound lowland was completely gone.

In 1962, Seattle opened the World's Fair with the completion of the space-age spire known as the Space Needle. There are pictures of me watching this construction, and I was fully aware when the magnitude 6.7 earthquake shook our home and tumbled our chimney on April 29, 1965. Seven

years later, my dad made me climb Mt. Rainer with him, and that brutal ascent remains the hardest thing that I have ever done. By the time I was thirteen, I had been exposed to the fact that the Seattle area was a place of geologic extremes, but I had not connected the dots and realized that

we lived in one of the most geologically active places in the world. That realization would not fully seat itself in my mind until Ray and I started working on this book, and I began thinking about how few Seattleites had realized that they live at ground zero on a dynamic planet.

The Seattle waterfront of the 1960s was dominated in my mind by two iconic establishments: Ivar's Acres of Clams and Ye Olde Curiosity Shop. Both were run by gregarious quirks. "Daddy" Stanley capitalized on the gold rush and opened the Curiosity Shop in 1899, taking advantage of the people who were returning from the goldfield, luring them in with souvenirs. Soon he was paying local Native Americans to carve totem poles. By 1909, his collection had grown so large that it was exhibited at the Alaska-Yukon Exposition and was later sold to George Gustav Heye, whose collection comprises the founding collection of the Smithsonian's National Museum of the American Indian. "Daddy" would buy artifacts, trinkets, and handiworks from visiting Native peoples, and the shop became a destination for all manner of curious visitors. In 1902, Chief Joseph, the famous Nez Perce, visited the shop, and, almost predictably, Teddy Roosevelt was there a few years later. Six decades later, I could regularly be seen spending my allowance there every weekend.

Ivar Haglund opened the seafood restaurant in 1946 and made a name for himself by cracking crab and clam jokes and sponsoring spectacular Fourth of July firework displays over Elliott Bay. In the 1970s, he became a local hero for sponsoring *Monty Python's Flying Circus* on the local public television channels.

Around the time I was ten, my dad bought me a boat. We lived in Beaux Arts Village, a curious little town on the shores of Lake Washington.

Founded in 1908, Beaux Arts styled itself as an artist colony just a short steamboat ride from Seattle. By 1970, that art dream was distant, and the colony had been engulfed by the surging growth of Bellevue. But Beaux Arts still had a beach and I had my boat. My boat was 12 feet long, green, wood, and heavy as a log. It was everything I could do to flip it over and slide it into the lake. Our neighbor, Harold Hanson, a Boeing engineer with a passion for old cars, had built the boat himself. He sold it to Dad for $10. The oars cost fifteen. For $25, my dad linked me forever to the water. My parents were remarkably lax with their supervision, and I was given leave to row about whenever and wherever I wanted.

The Lake Washington of the early 1970s was just beginning to recover from a phase of pollution, and my friends and I would fish for perch and dive for crawfish. There were sockeye salmon that would enter Lake Washington from Puget Sound through the fish ladder at the government locks in Ballard, and then make their way down the lake to the Cedar River where they would spawn. We learned that trolling deep and slow with just the oars was the right way to get one of those magnificent fish to strike a lure. The first thing I ever cooked was a freshly caught sockeye over an open fire on that beach.

So it came to be that by the time I was seventeen, my interests firmly were bent toward fossils, fishing, driving, drawing, carving, volcanoes, coastlines, tide pools, forests, boats, and hiking. I got summer jobs working for the Pacific Fish Company and as a longshoreman unloading the coastal freighters that serviced the little fishing towns of the Aleutian Islands. I longed to go to Alaska.

Ray moved to Seattle in 1977, within a month of when Wes and I found the fossils at Republic. He moved there on whim, started slinging beers at the Aurora Tavern, selling art at the Pike's Place Market, and working at Allied Arts in Pioneer Square. He joined a band called the Bones of Contention. In 1979, he headed over the mountains to Pullman where he got a master of fine arts – and another band. His first job was teaching art to bored Coast Guard personnel in Port Clarence on the Bering Sea. Alaska had claimed another Seattleite. In 1983, he moved to Ketchikan, where he lives today. The arc of his life shifted strongly from art and music to art, music, fish, and T-shirts. He became a T-shirt mogul famous to anyone who fished the coast from Oregon to Alaska.

Opposite: Festooned with walrus heads, the Arctic Building in Seattle was built in 1916 by men who returned from the Klondike Gold Rush.

Above: Built in 1914, Seattle's Smith Tower was once the tallest building west of the Mississippi.

For all these reasons, Seattle feels like home to both Ray and me. It was here that I first met him in 1993 at an exhibit at the Burke Museum called *Planet Ocean*. Sixteen years later, we were back at the Burke, working together on an exhibit called *Cruisin' the Fossil Freeway* and sneaking out to have a good look at the local fossils.

Seattle is a rainy place and rainy places are full of trees. Trees cover the ground and obscure the geology, so there are only a few places to find bedrock in a place like Seattle. Those places are roadcuts, construction sites, riverbeds, and rocky beaches. Like any city of note, Seattle has its construction fossils, like the Jefferson ground sloth that was found during the excavation for a light pole on the runway at Sea-Tac airport in 1962 or the mammoth tusk found in 2015 in downtown Seattle.

Ray knew a geologist named David Montgomery who studied soil and played in a band called Big Dirt. David is a great writer as well, and he

Upper: Ray and David Montgomery whack their way through the ferns.

Middle: On the banks of an Ice Age salmon-filled lake.

Lower: A fossil sockeye salmon.

DOWN AT THE SOCKEYE HOLE, ONE MILLION YEARS AGO.

had written a sweet book about salmon that Ray illustrated. You wouldn't think that there would be much connection between salmon and dirt, but there is. Salmon are anadromous, which means they spend their adult lives in the open ocean but spawn in the headwaters of freshwater rivers and streams. In the Northwest, we call these rivers "salmon streams." Most salmon die after they spawn, and they become food for bears, bobcats, wolves, eagles, and many other animals. And we all know what a bear does in the woods. Salmon bodies are rich in nutrients, and the animals that eat salmon spread those nutrients across the forest floor where they get worked into the soil and eventually becomes the stuff of trees.

David works at the Quaternary Research Center at the University of Washington, which is a fancy way of saying he studies the geology of the Ice Age. He had found a really odd fossil site at the south end of Puget Sound, and we talked him into taking us there. We met him in Olympia and drove around the south end of the sound to the valley of the Skokomish River.

Since most of the land around Puget Sound was logged between 1860 and 1960, most of the modern forest is second- or third-growth timber, and a lot of land is owned by timber companies who are sitting around waiting for their next crop of trees to grow back. For this reason, the foothills of the Olympics and Cascades are cut with a maze of dirt logging roads that wind their way through brushy young forests.

Dave led us through a locked forest company gate and down an old, overgrown logging road through a tangle of small alder trees and blackberry brambles. Eventually we popped out onto the bank of a beautiful little salmon stream. It was the kind of place that made me wish I had a fly rod with me. It was a crystal-clear creek with a gravelly bottom, but as soon as I got to the river's edge, my geologist's eyes spotted something very unusual. The bank of the river was composed of clay that had a very distinctive type of layer known as a varve. Varves form at the bottom of lakes that form at the end of a glacier. Right away, I realized that I was in a river valley that used to be a glacial lake.

This all makes sense if you think about Harlen Bretz's ice sheet filling Puget Sound. At a time like that, the rivers (and indeed glaciers) flowing down valleys from the mountains collided with the massive ice sheet that filled Puget Sound. These collisions would block the mouths of the river valleys and cause them to fill up with water and become lakes. David confirmed that we were looking at a 1-million-year-old glacial lake. Then he told me to go ahead and split open some of the clay. Using my Swiss Army knife, I started splitting the soft clay and immediately started seeing fins, scales, and finally a tail. After a few minutes I had exposed a beautiful fossil salmon. And this wasn't just a salmon. It was a sockeye salmon, just like the ones I used to catch in Lake Washington.

Here was a million-year-old glacial lake full of spawned-out sockeye salmon that would have had to make their way from the open ocean through an ice-plugged Puget Sound and up an ice-choked river past mammoths and saber-toothed cats before spawning, dying, and getting buried. For the first time in my life, I had a visceral understanding that I had grown up in a postglacial landscape.

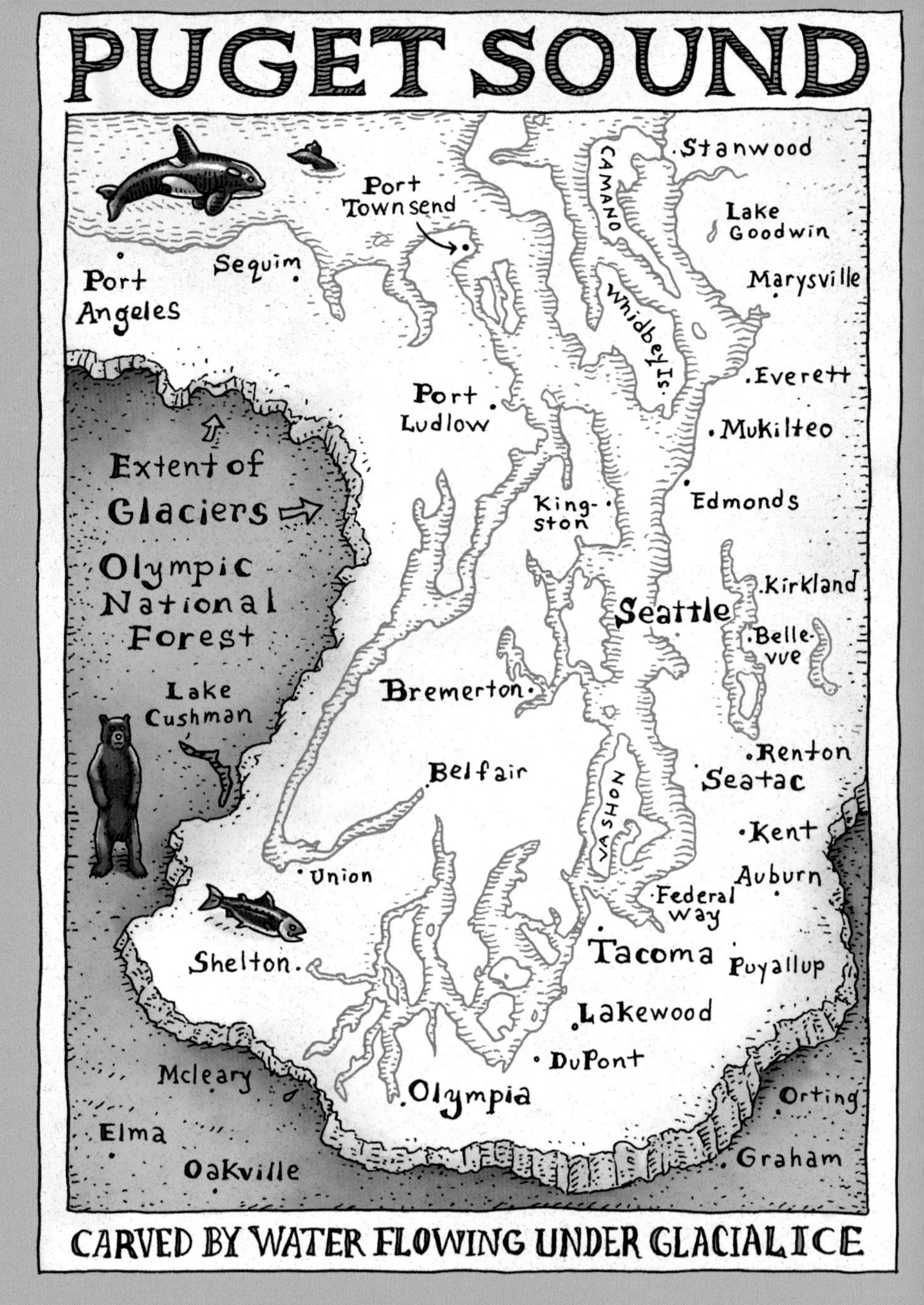


David drove us back out to the highway and we parted ways. Ray and I headed north along the shores of Hood Canal. In truth, it is a misnomer to call this a canal because that implies it was dug by somebody. In fact, the name does appear to be a mistake, as it was originally noted as a channel in the logbook of Captain George Vancouver, who surveyed it in 1792. Hood Canal is a long, narrow body of saltwater that stretches 50 miles from Puget

Sound to the southwest before doubling back on itself to form a giant fishhook. The "canal" is a little more than a mile wide and almost 200 feet deep.

It was a gorgeous sunny afternoon as we drove past oyster farms and giant blackberry bushes heavy with berries. With my new Puget Sound–glacial perspective, I pondered the process that could have formed this curiously shaped canal. I came up blank. I asked Ray what he thought, and he gave me the simple answer that you see on Wikipedia: The glaciers caused it. I knew that. What I wanted to know was what was it that the glaciers did that caused all of those long channels that are the signature topography of the Puget Sound lowlands. Now that I had a simple question that I could not answer, I became obsessed with it.

Over the next few weeks, I did what scientists do to answer questions. I searched the published scientific literature about the geology of the basin and I called some experts. Surely someone must know, and it turns out that they had been asking the same question for a long time. I learned that in 1884, John Strong Newberry – who was a paleobotanist, by the way – published a paper called "On the Origin of the Fiords of Puget Sound." Bretz also puzzled over the pattern and began to unravel the answer.

As we drove north along the western shore of Hood Canal, we were also traversing the eastern margin of the Olympic Peninsula, one of the more majestic and mystical places on the planet. The peninsula is cored by the jagged Olympic Range, which is composed of a giant horseshoe-shaped fold of baked and tortured seafloor that was shoved up onto the edge of the North American continent. The range creates a major impediment to the eastbound weather off the North Pacific, and the west side of the peninsula has some of the highest rainfall levels in the world. Some spots see more than 200 inches a year, and this ample moisture is the reason for the incredible rain forests. The coast is a jumble of jagged rocky headlands interspersed with rocky beaches covered with Japanese flotsam. It was on these beaches that I saw my first tide pools and crashing ocean waves.

## Carving the Sound

The Puget Sound Basin existed as a lowland before the glaciers arrived from Canada. The rivers would have flowed into the basin and generally to the north. When the south-moving glacier advanced into the basin, it dammed these rivers, creating a giant lake. As the glaciers continued south and filled the basin, meltwater from the glacier would have flowed to the south over the divide by Olympia and into the Chehalis River, where it could flow into the ocean near Aberdeen. It appears that meltwater at the bottom of the glacier carved river channels under the ice. When the glacier retreated, it exposed these channels and flooded them with fresh water, forming a huge lake at the south end of the glacier – one that would have stretched all the way from Olympia to Everett. Once the glacier retreated beyond the Juan de Fuca Strait, the freshwater lake emptied into the ocean, the salt water surged in, and Puget Sound was formed. To make things even more complicated, it appears that this entire process happened more than once. It's no wonder that people just say that the glaciers did it.

OYSTER BAR
So an oyster bear walks into an oyster bar....


Neah Bay is located just east of Cape Flattery, and the northern coast of the peninsula forms the southern shore of the Juan de Fuca Strait. The strait stretches 90 miles to the mouth of Puget Sound, and a number of spots along this coastline have given up Eocene, Oligocene, and Miocene marine fossils similar to those found on the Oregon coast. Facing the strait, this coastline is not bashed by rolling ocean waves, but its rocks are hard, its tides tricky, and the weather is often rotten. It was here that Doug Emlong found another skull of the animal known as *Kolponomos*, the oyster bear (even though it is not found with oysters) and a superb jaw from a primitive desmostylian called *Behemotops*.

Cape Flattery, located on the Makah Indian Reservation, is the northwesternmost corner of the lower forty-eight states. Just to the south at a place called Ozette, a Makah village was buried in a landslide about 400 years ago. Coastal erosion exposed the edge of this village in 1970, and delicate artifacts began to erode onto the beach. The state partnered with the Makah Tribe to undertake an eleven-year excavation that yielded thousands of artifacts, and in 1979, the state opened the Makah Cultural and Research Center in Neah Bay. The Makah are a whaling people who hunted baleen whales from cedar canoes. In 1999, the tribe revived this tradition and killed a gray whale. The rocks of the Olympic Peninsula are fossiliferous, and in places they contain the remains of some of the world's earliest known whales, preserved alongside the remains of other denizens of the ancient Pacific.

Coastal areas that have fossils often have people who are obsessed with those fossils. Growing up in Seattle and haunting the halls of the Burke Museum, I heard about these people and tried to meet them. Over time, I met Bill Buchanan of Clallam Bay, Bill Brandt of West Seattle, and Ross Berglund of Bainbridge Island. All three were old men when I was young, and they all generously showed me their fossil collections and gave me tips on how to find my own. And all three shared their finds with scientists who would add their discoveries to the scientific literature and museums who accepted their generous donations.

As the old-timers passed away, they have been followed by younger collectors. Over the years, I began to hear of a prolific collector named Jim Goedert. Jim, whose day job is working as a railroad lineman, started collecting in the 1980s and learned some of his skills from Buchanan and Berglund. In 2008, a collector named John Cornish found a fossil whale snout sticking out of a rock wall in a quarry west of Port Angeles. He called Jim Goedert, who arranged for it to be collected by the Burke Museum. The specimens turned out to be a surprisingly complete baleen whale that is now on display at the museum. The amount of time it takes to hike the beaches in search of rare finds has made coastal paleontology the realm of the obsessed amateur. In Goedert's case, this dedication has made him a scientist. When I finally met him for the first time in the waning days of 2016, I was startled to learn that he is only four years older than I am, and he has published more than 100 scientific papers on the paleontology of western Washington.

As a teenager, I had avidly watched the coverage of the Ozette excavation. And I had heard rumors of fossil whales and odd beach bears. I convinced older kids to borrow their parents' cars and take me on fossil hunting jaunts. We would stroll down the beaches, watching for the incoming tides, and find what we could find.

On one beach, I found concretions that were so incredibly hard I could not break them open with my 4-pound sledgehammer. I came back with an 8-pound sledge, and that made a difference. Soon I was smashing the orange-sized rock balls and watching them pop in half. Many of them had beautiful little fossil crabs. Of course, hitting a round rock with a big hammer on a slippery beach in the rain is not the safest of teenaged sports. A glancing blow could send the rock flying like a golf ball. I caught a few of those in

Left: *Galeodea apta*, an Eocene snail from the Olympic Peninsula.

Above: Concretions on the beach at Marrowstone Island.

my shins. Too hard of a blow could smash the concretion into shards.

One day, I was hunting by myself and the rocks simply weren't breaking. I could literally bounce the sledge off the recalcitrant rocks. Eventually, I came upon a larger concretion about the size of a big grapefruit. Since this was about twice the size of the average concretion, I judged that I needed about twice the force. I took a big over-the-head swing and gave it all I had. Much to my surprise, the concretion exploded open and shattered into a dozen pieces. I saw to my horror that it had broken easily because the concretion was largely filled with a large pearly nautilus shell that had taken the brunt of the blow. I spent the next hour carefully collecting all of the shards. It took me a couple of weeks but I finally got Humpty-Dumpty back together again.

Remembering the spot and noting the tide, I took Ray to this beach, and we arrived in the early evening about an hour before sunset. It was the end of a calm day, the sea was flat as glass, and the tide was going out. We didn't have a big sledgehammer, so I knew it was unlikely that we would find anything. It had been more than thirty years since I had visited this beach, but it was much as I remembered. Fossils were everywhere – sticking out of the bank, eroded on the shore, or poking out of concretions. We found clams, tusk shells, crab claws, and pieces of fossil wood pockmarked with round dots of sand that signaled the wood had been burrowed by *Teredo* clams before it had been buried. Even though the fossils were 30 million years old, they told a story of life on the coast. This is the same story that the place itself tells today.

Miocene fossil starfish (*Zoroaster*) from Ozette.

After the sunset, we drove into Port Townsend to meet some friends. They were down on the waterfront and we joined them for a beer. A couple of park rangers from the Olympic National Park joined us, and we got to talking about fossils. The rangers had been hiking the beach near Ozette and had found a fossil starfish on an outcrop. They showed us some pictures, and it was clear that the unusual fossil would soon be destroyed by the violent waves. I told them how to use a diamond-bladed rock saw to extract the fossil. Later we learned that they had been successful, and the fossil was on display at the Burke Museum.

The northern Olympic Peninsula has also produced its share of Ice Age fossils. On August 8, 1977, near the town of Sequim, Emanuel "Manny" Manis was excavating a pond on his property when his backhoe uncovered two ivory tusks of an American mastodon. The discovery caused a huge stir, and more than 50,000 people visited the farm to have a look. Years later, closer inspection of the skeleton showed that a bone projectile point was embedded in one of the animal's ribs. Radiocarbon dating of the skeleton showed that it was 13,800 years old, making it some of the earliest evidence for people in North America.

Many of the islands around Port Townsend have a bedrock core and a generous coating of glacial sediments. The next morning, we joined some friends for a walk along the rocky shore of Marrowstone Island, and we found a number of spots where big soccer ball–sized concretions were emerging from the rocky outcrops along the beach. But we weren't interested in the bedrock, we were interested in seeing something that one of our friends had found in the glacial sediments. After a few miles, he pointed to the cliff, and we walked over to look at a strange white object about 8 inches in diameter. It had growth rings, but it wasn't a tree. After a close look, it was clear that this was the cross section of a mammoth tusk. We were on the north end of Puget Sound looking at another bit of evidence from the icy world.

Later that day, we took the Washington State Ferry from Port Townsend on a short 6-mile run across the sound to a place called Keystone. While crossing, we avoided a number of small fishing boats that were trolling for salmon.

Once a mainstay of the state economy, commercial salmon fishing had declined in tandem with the logging and urban development of Puget Sound that had harmed, polluted, or destroyed salmon spawning streams. In Puget Sound, the condition of the land has a direct impact on the

Upper right: Clare Manis Hatler with a photo of her late husband, Manny, and his mastodon tusks.

Right: Cross section of a mammoth tusk in a sea cliff on Puget Sound.

FISHES OF THE SALISH SEA

health of the marine ecosystems. Native American tribes still fish for salmon using gill nets in the rivers, and there is still a sports fishery for salmon – but the salmon runs are a mere shadow of what they were a hundred years ago.

In recent years, there has been a strong push to restore marine ecosystems in Puget Sound, and this has led to the realization that Puget Sound is part of a larger series of inland coastal waterways that includes the region between Vancouver Island and the British Columbia mainland. In 1988, a marine biologist named Bert Webber proposed the name "Salish Sea" for the area, and that name has stuck. In 2015, ichthyologists Ted Pietsch and James Orr did a survey of all known fish species in the Salish Sea and documented a total of 253.

RATFISH WAITING PATIENTLY FOR SEATTLE TO GO AWAY.
AQUARIUM


## Introduction to the Ratfish Empire

I can remember my first ratfish like it was yesterday. I was in a small boat with my dad fishing for salmon in Shilshole Bay near Ballard. Something took my bait. It definitely was not a salmon – it was something much milder. When I reeled it in I was amazed with what I saw. If a fish could have buck teeth and a rat tail, this fish did. It also had luminous green eyes, and its head shimmered in a rainbow of colors. It was simultaneously the most beautiful and most ugly fish I had even seen. I had heard that anglers rendered ratfish oil to lubricate their reels, but my reel seemed fine. I tossed it back.

Ray had known Ted for more than thirty years, and together they dreamed of illustrating the entire fish fauna of the Salish Sea. With Ray, dreams often end up as paintings, and in 2011 he completed a Salish Sea masterpiece, a 7-by-15-foot acrylic on canvas mural showing ninety-nine species of Salish Sea fish (pages 154–155). The finished painting hangs on the wall of the University of Washington Fisheries Sciences building on Portage Bay. The number of known fish species continues to grow because fish scientists continue to explore the Salish Sea, but this is not the whole story. The thing that had changed is that the relative abundance of different species has dramatically shifted, and the absolute numbers of the five different salmon species has dropped dramatically. The salmon have been replaced by spotted ratfish (*Hydrolagus colliei*), and some estimates suggest that perhaps as much as 75 percent of the fish biomass in the Puget Sound is this one species. It is for this reason that Ray and I call the Salish Sea by a different name: the Ratfish Empire. In actual fact, since the range of the spotted ratfish stretches from Alaska to Baja, and the ratfish lineage reaches back to the Paleozoic, this entire book takes place under the jurisdiction of ratfish.

For years, Ray has been leading a band called the Ratfish Wranglers. He calls himself Ratfish Ray, and there is even a species of ratfish named after him (yep, *Hydrolagus trolli*). Ray loves ratfish. But unless you are an angler or someone who frequents aquariums in the Northwest, chances are that you have never seen a ratfish.

As we crossed the channel, we also talked about orcas. A few months before, I had taken the Washington State Ferry from Edmonds to Kingston and had seen a huge pod of orcas swimming on both sides of the boat. Despite all of my time on the water in Puget Sound, this was the first time I had seen orcas from a ferryboat and it blew my mind. *Orca orcinus* is the largest of the dolphins, with males up to 25 feet long and weighing up to 6 tons. Their common name is the killer whale, and it is not a misnomer as orcas do kill and eat dolphins, porpoises, great white sharks, seals, otters, sea lions, walrus, narwhals, salmon, herring, elephant seals, seabirds, huge baleen whales, and just about anything else they want. Not all orcas eat all of these prey items, and many pods specialize on certain ones, honing their hunting skills and training their young in the family business. They are smart, fast, and work as a team, and they are gorgeous. The resident population of orcas in the southern part of Puget Sound has about eighty animals that are part of three different pods. Known as pods J, K, and L, these orcas focus their feeding on salmon.

Historically, orcas have ranged throughout the world's oceans, but nowhere are they so iconic as in the Pacific Northwest. It is impossible to visit Seattle, Vancouver, or Alaska and not see paintings, photographs, and sculptures of orcas. They are also well represented in the iconography and art of the Native peoples of the West Coast. Clearly these animals have made an impression on all humans that have encountered them.

Because of their beauty, size, and intelligence, orcas became the star attraction in marine parks around the world. At the same time, scientific research on the ecology of orcas of the Salish Sea began to reveal the ecology and community structure of the whales. Scientists working with volunteers began to photograph individual orcas and, in time, created a catalog of all of the residents. Underwater recordings of their calls demonstrated that different pods had different dialects. Using recordings of Namu, scientists were able to identify Namu's pod and his surviving relatives. Protests finally led to the end of capturing wild orcas in the Salish Sea, but many remained in captivity around the world.

## Namu, the Captive Orca

In June 1965, a 24-foot-long male orca got tangled in a fishing net near Namu, British Columbia. The fishermen, sensing an opportunity, decided to see if they could sell him. Ted Griffin, the owner of the short-lived Seattle Marine Aquarium (not to be confused with today's Seattle Aquarium, which opened in 1977) paid $8,000 to the fishermen, named the whale Namu, and then had to figure out how to get him to Seattle.

At the time, my dad worked out at Harry Sweatnam's downtown gym with a local AM radio personality named Robert E. Lee Hardwick. Hardwick was known for outrageous pranks like swimming the 14-mile ferry route from Bremerton to Seattle or jet skiing the 740 miles from Seattle to Ketchikan. Since Dad knew Hardwick, our family had a ringside seat for what happened next.

Griffin and Hardwick cooked up a plan to build a floating pen to contain Namu and to tow him the more than 450 miles back to Seattle. Hardwick piloted the tugboat that towed the pen and set out for Seattle. On the fourth day of the trip, a pod of thirty to forty orcas started following the pen and stayed with it for a few hours. Three whales, likely close relatives of the captive, followed the tug and its woeful cargo for 150 miles before giving up.

Once back in Seattle, the whale became a huge attraction. I was almost five years old at the time and can clearly remember peering over the edge of the dock into the giant pen. Countering the prevailing wisdom (and common sense) of the time, both Griffin and Hardwick started swimming with Namu. For Hardwick, it was probably just another stunt. For Griffin, it became a way of life. Namu died in his pen just a year after his capture, and Griffin would go on to capture and train dozens of orcas, selling them to aquariums for $20,000 a pop.

In 1993, *Free Willy,* a film about a relationship between a boy and a captive orca, accelerated growing sentiment that captive orcas should be released. Keiko, the *Free Willy* orca, was moved from his tank in Mexico to the Oregon Coast Aquarium in Newport to be staged for release. In 1998, Ray had a show at the aquarium, and he and I both spent a lot of face time with Keiko. Eventually, Keiko was released from a pen to the open ocean off of southern Iceland. In 2013, the documentary film *Blackfish* made a powerful case for the release of all captive orcas.

On this day, we saw no orcas on the passage from Port Townsend, and the ferryboat landed at Keystone. We drove north along Whidbey Island to the bridge at Deception Pass and back onto the mainland. From there, we continued toward the delta of the Skagit River and through the small town of La Conner. We had one last stop to make before heading back to Seattle.

The outside patio of the Burke Museum contains a number of local rocks and fossils that are embedded in concrete plinths. I have known these monuments to local geology for my whole life, and one by one, I was figuring out where they came from and what they meant. These rocks have modest interpretive signs, and I always wonder what people think when they see them. A couple of the rocks contain the clear impression of palm fronds, not a typical plant of the modern Pacific Northwest. It turns out that these rocks were collected south of Bellingham in a geological formation known as the Chuckanut. The road from La Conner to Bellingham crosses the flat delta before climbing up along a rocky promontory and winding along the face of a steep wall for about 8 miles. The road has beautiful views to the west, with Guemes and Lummi Islands in the foreground and the San Juan Islands in the distance. It is one of the most scenic roads in Washington and a lovely afternoon drive.

The Chuckanut Formation is a mile-thick layer of sandstone and mudstone that was deposited about 50 million years ago during a time of intense global warmth in the early Eocene. The formation weathers rapidly and is normally covered by thick forest as is the case for much of the drive. For the last thirty years, George Mustoe at Western Washington University in Bellingham has been quietly collecting fossil leaves and footprints in the Chuckanut Formation.

What he has found por-

George Mustoe and a fossil palm frond from the Chuckanut Formation.

trays a radically different Washington State from what is here today. He has found dozens of species of fossil leaves that have the characteristics of modern tropical rain forests. These include big, broad, oval leaves with elongated tips useful for wicking off excess water, the leaves of cycads, and abundant palm fronds. Mixed in with these leaves are others that are similar to those we discovered in Republic.

If you connect the dots between Eocene Republic and Eocene Bellingham, it looks like the Washington of 50 million years ago was a place with enough topography to support cool uplands and warm subtropical lowlands. This is similar to the kind of climate range that you might find today between the mountains and coast of Georgia.

Ray had been on Chuckanut Drive before, but he had missed the best bit. I pulled over at one of the many scenic overlooks. It was a stunning evening, and the sun was about to set over the San Juan Islands. I had not told him why we had come to this spot, so his gaze bent toward the scenery of the west. This is the same mistake that

Above right: Russ Crane with a palm frond near Chuckanut Drive.

Right: Bart Weis and a large rain-forest leaf from the Chuckanut Formation near Bellingham.

everyone makes when they stop on Chuckanut Drive. After letting him enjoy the view for a few minutes I said, "Look east." He turned around and looked across the road at a big exposure of Chuckanut sandstone. There on the face of the cliff, high above the road, was a wall of rock completely covered with fossil palm fronds.

We had brought oysters with us from Port Townsend so we drove down to the beach and built a little fire to roast them. As the sun set, we pondered the mysteries of the Salish Sea and the fact that 3,000-foot-thick ice sheets had once occupied the same ground as palm forests and killer whales.

A FOSSIL MAP OF
BRITISH COLUMBIA AND ALBERTA
Yakutat
Paleo-Crab
U.S.A.
CANADA
Mountain Goat
Carcross
Watson
Jake's Corner
Teslin
Alaska Highway
Beringia or BUST!
Watson Lake
YUKON
NORTHWEST TERRITORIES
Fort Liard
Trout Lake
Rhizodontid
Enterprise
Lobefin Trackway...?
Grumber
Skagway
Klukwan
Haines
Glacier Bay Nat. Park
Gustavus
Juneau
Elfin Cove
Humpback Whale
fossil palm
Pelican
Chichagof
Admiralty
Hoonah
Angoon
Sitka
Baranof Is.
Near Kake
Coryphodon tracks..?
Kake
Kuiu Is.
Petersburg
Devil's Thumb
Thallatosaur near Kake Alaska
Wrangell
Teepee
Engineer
Atlin Lake
Surprise!
Adrianites
Swift River
Squirrel
Moose
Jade City
Nahlin
Good Hope Lake
Cambrian Sponge Fossils
Red Fox
Fireside
Coal River
Helicoprion the Buzz Saw Shark
BRITISH COLUMBIA
Nelson Forks
Clarata
Internites Ammonite
Helmeted Musk Ox Bootherium
Dease Lake
Stikine River
Bighorn Sheep
Muncho Lake
Toad River
Summit Lake
Muskwa
Shonisaurus sikanniensis Nearly 70 ft. long!
Fontas
Fossil Ficus Leaf
Mountain Lion
Woolly Mammoth
World's Largest known marine reptile
Lower Cretaceous Plants - Bowser Lake
Cambrian Sponges
Euflemingites
Gulosaurus helmi
Sikanni Chief
Triassic Beasties
Acrodus shark
Ostrea
Brown Bear
Ursus arctos
Stephanoceras Middle Jurassic
Dino Tracks
Fort St. John
Hudsons Hope
Williston Lake
Taylor
Silurian Brachiopod
Prince of Wales
Craig
Ketchikan
Unuk River
Hyder
Stewart
Sustut Dinosaur Cerapoda incertae sedis Still a mystery
Swan Lake Prov. Park
Ordovician Graptolites
Malayaspira
Rebellatrix the "rebel" Coelacanth
from Wapiti Lake
Dwarf species
Dawson's Caribou Extinct 1908
Cranberry Junction
Nisga'a Lava Bed Prov. Park
Kispiox
Bison latifrons
Lambeosaur
Dawson Creek
Tupper
Mackenzie
Peace Region Paleontology Centre
Misty Fjords
Kitwanga
New Hazelton
Smithers
Telkwa
Ankylosaur
Tracks!
Tumbler Ridge Museum
Wolf
Usk
Terrace
Topley Landing
Houston
Eocene Plants
Jurassic Radiolaria
Prince Rupert
Skeena River
Kitimat
Listracanthus
Bear Lake
Wapiti Lake
Triassic Fish!
Eocene Tapir & Hedgehog
Burns Lake
Endako
Fraser Lake
Prince George
Saurichthys
Early Jurassic Ichthyosaurus
Badouxia Ammonite
Queen Charlotte
Tlell
Skidegate
Sandspit
Inoceramus
Hartley Bay
Heptodon
Kermano
Silvacola
Megalonyx jeffersonii Ground Sloth
Red Rock
Dubariceras freboldi
PACIFIC OCEAN
Haida Gwaii
Moresby Camp
Triassic Radiolaria
Spirit Bear
A.K.A. Kermode Bear
Sockeye
Cinema
Troll Ski Resort
Bobasatrania
Burgessia
Valemount
Haplophrentis
Jasper
Metaderoceras
Amyzon Eocene Sucker
Nazko
Quesnel
Short Faced Bear Vancouver Island
Bella Coola
Terror of the Cambrian
Anomalocaris canadensis from the Burgess Shale
Psuedhelicoceras Ammonite
Campbell Island
Bella Bella
Vampire Squid Cretaceous
Namu
Knot Lakes
Chilanko Forks
Redstone
Williams Lake
Horsefly
150 Mile House
Sassafras
Enchodus
Edaphodon fossil ratfish
Pleistocene Steller Sea Lion - near Courtenay
Tatlayoko Lake
Arctodus
Eocene Lakebed Fossils
Clearwater
Wiwaxia corrugata
Trilobite beds!
Olenoides
Yoho Nat. Park
Field
Burgess Shale
Pseudoxybeloceras nanaimoense
six gill shark tooth
Glyptoxoceras
from Vancouver Island
Kourisodon puntledgensis Mosasaur
Kingcome Inlet
Port Hardy
Port McNeill
Alert Bay
Bute Inlet
Chasm
Clinton
Cache Creek
Hallucigenia sparsa
Mastodon
Paleo-Red Salmon
Kamloops
Yawunik kootenayi
Armstrong
Metaspriggina walcotti
Nakusp
from Nanaimo
Albertonia cupidinia
from Wapiti Lake
"Rosie" the Pleistocene Walrus
from Qualicum Beach
Fossil Fern
Kyuquot
Campbell River
Columbian Mammoth
Lytton
Pleistocene Salmon
Oyama
Opabinia regalis
from Burgess Shale
Puntledge River Elasmosaur
A very big ammonite
Courtenay & District Museum
Powell River
Vancouver Aquarium
Gold River
From Alberta of course!
Albertosaurus
Didymoceras hornbyense Heteromorph Ammonite
VANCOUVER ISLAND
Qualicum Beach Historical & Museum Society
Port Alberni
Nanaimo
Science World
Vancouver
Surrey
Hope
fossil flower
Princeton
Eosalmo the "dawn" salmon
Metasequoia occidentalis
Grand Forks
Salmo
Chilliwack
Abbotsford
Tofino
Ucluelet
Clo-oose
Port Renfrew
Cornwallius sookensis
A cool 'Desmo'
Royal B.C. Museum
Victoria
Sooke
Pacific Museum of the Earth
Woodland Muskox near Victoria
Sedro Woolley
Oroville
WASHINGTON
Republic
Stonerose Interpretive Center
ALBERTA
Maiaspondylus lindoei
"Momma" Ichthyosaur
mom!
Rainbow Lake
High Level
John D'Or Prairie
"Float & Bloat" Nodosaurid
found near Ft. McMurray
Pleistocene Elk
Hotchkiss
Manning
Magnolia
from Watino
Clear Prairie
Gar
Dixonville
Pachyrhinosaurus
Peace River
Grimshaw
Xiphactinus
Mckennan
Pliopl[atecarpus]
Guy
Pipestone Creek Bonebed
Wembley
Grande Prairie
Philip J. Currie Dinosaur Museum
Kleskunsaurus grandeprairiensis
Dino Tracks
Ficus
Birgeria
Gabriellus lanceatus
Mickwitzia
Banff National Park
Necrocaris
Alokist[odon]

# BRITISH COLUMBIA

# 9 AMMONITES OF THE SALISH SEA

The San Juan Islands of Washington State and the Gulf Islands of British Columbia form a cozy archipelago split by the international border. The rocks here are a smashed mishmash of largely Cretaceous sedimentary rocks known as the Nanaimo Group that was later overrun, gouged, and polished by the ice sheet from the north. These islands are the mellow cruising grounds of the fleets of pleasure boats that stock the marinas of the Salish Sea. Sheltered from rain by the Olympic Mountain rain shadow, they are dry and mild. Pods of orcas favor the west side of San Juan Island. All in all, these idyllic islands are the kind of place you imagine you might retire.

As a teenager, I crewed on sailboats around the islands and worked at a salmon cannery in Friday Harbor. I have very sleepy but strong memories of spending long nights up to my waist in concrete holding tanks full of pink salmon. This species of salmon are also known as "humpies" because

Long-necked elasmosaurs, *Saurodon* fishes, ammonites, baculites, and a vampire squid cruise the waters of the Cretaceous Salish Sea.

the males develop exaggerated humpbacks (and hooked jaws) during spawning. One of Ray's most popular T-shirt designs was one entitled "Humpies from Hell" and was based on the general hatred of all cannery workers for the pink salmon because of its soft flesh and propensity to spoil. Having lived my own Humpy Hell, this shirt was the first image that made me aware of Ray Troll. The pink salmon commercial fishery is now extinct in the San Juan Islands, and the cannery is just another abandoned building. Given their geological history, the San Juan Islands also hold the evidence of other extinctions.

One island had a special place in my memory because of a specimen case at the Burke Museum that was full of pearly shelled ammonites and baculites. The labels said Sucia Island, and I soon learned that the Burke Museum had launched a series of trips to this oddly shaped little island to extract marine fossils from the Nanaimo Group. Eventually I visited the island myself, only to learn that it was a state park and no fossils could be collected without a permit. I walked the wave-cut terraces and saw a variety of fossils poking out of the crumbly slopes.

In January 2014, I came to Seattle to present a lecture about the dangerous geology of the Pacific Northwest. It was a Rip Van Winkle talk because so much had changed in the thirty-six years since I had moved away in 1978. What had changed was a raft of new discoveries about faults, earthquakes, tsunamis, landslides, and lahars.

One thing that had not changed was that there were still no dinosaur fossils from Washington.

The Vancouver Island Cretaceous Sea

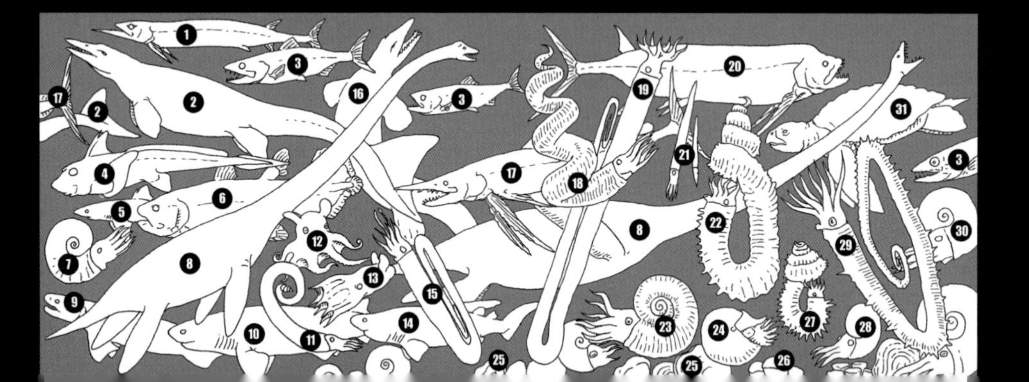

1. *Saurodon leanus* – marine fish
2. *Kourisodon puntledgensis* – mosasaur
3. *Enchodus pterosus* – marine fish
4. *Edaphodon hesperis* – ratfish, AKA chimaera
5. *Squalus* sp. – dogfish shark
6. *Megalocoelacanthus* – giant coelacanth
7. *Gaudryceras striatum* – ammonite
8. Puntledge River elasmosaur
9. *Chlamydoselachus* sp. – frilled shark
10. *Hexanchus griseus* – bluntnose sixgilled shark
11. *Glyptoxoceras subcompressum* – heteromorph ammonite
12. *Paleocirroteuthis haggarti* – dumbo octopod
13. *Nanaimoteuthis jeletzkyi* – early vampire squid
14. *Echinorhinus* sp. – bramble shark
15. *Polyptychoceras vancouverense* – heteromorph ammonite
16. *Tylosaurus* sp. – mosasaur
17. *Protosphyraena perniciosa* – marine fish
18. *Hyphantoceras* – heteromorph ammonite
19. *Diplomoceras* – heteromorph ammonite
20. *Xiphactinus audax* – marine fish
21. *Baculites inornatus* – baculite
22. *Bostrychoceras elongatum* – heteromorph ammonite
23. *Canadoceras yokoyama* – ammonite
24. *Nautilus campbelli* – nautilus
25. *Inoceramus vancouverensis* – bivalve
26. Rudist clam
27. *Nostoceras hornbyense* – heteromorph ammonite
28. *Gaudryceras striatum* – ammonite
29. *Pseudoxybeloceras nanaimoense* – heteromorph ammonite
30. *Gaudryceras striatum* – ammonite
31. *Desmatochelys padillai* – sea turtle

I had recently surveyed the fifty states and learned that only thirty-six had yielded dinosaurs. The unlucky fourteen were places like Maine, Vermont, Hawaii, and Rhode Island, where there was little to no rock from the Triassic, Jurassic, or Cretaceous periods, the time when dinosaurs walked the land. There were a few states with Cretaceous rocks but no dinosaurs. Washington was one of those states.

In my talk, I made fun of the geologists in the room and challenged them to work harder to resolve the Washington State dinosaur deficit. After the talk, and much to my surprise, a University of Washington professor came up to me and quietly said, "I can't tell you anything definitive yet, but we have just found the state's first dinosaur." I had to wait fourteen months until Brandon Peecook and Christian Sidor published a paper describing a partial femur of a tyrannosaurid dinosaur that had been discovered on Sucia Island by none other than Jim Goedert.

Jim had been visiting Sucia with David Starr, a Seattle-based stockbroker and fossil enthusiast who had obtained a permit from the park to collect fossils. On one trip, Jim found a loose piece of broken bone but left it on the beach because it had no diagnostic features and was thus worthless to science. He thought that it might be the remains of a marine reptile such as a mosasaur, which is what you would expect to find in a geologic formation full of ammonites. On a subsequent trip, Jim found a larger bone sticking out of the bedrock and realized that the earlier fragment must have broken away from the larger piece still embedded in the rock. He searched the beach and relocated the previously discarded chunk and found that it was part of the larger bone. Now that he had something identifiable, he called the Burke who excavated the bone and realized it was from a dinosaur.

Dinosaur paleontologists have a phrase for the phenomenon of land fossils in marine deposits. They call them "bloat-and-float" fossils. Imagine that an animal dies along the banks of a coastal river and has a chance to bloat with the postmortem gases of decay. If this fetid carcass then gets carried downstream, it is possible that it will drift out to sea. Eventually it will lose its buoyancy and sink to the seafloor, where it is buried in the mud alongside the remains of marine animals from the water column and the seafloor. For this reason, there are several dinosaurs known from marine deposits around the world. Jim had found Washington's first dinosaur, and it was one bone from a bloat-and-float corpse.

Left: Strolling Sucia Island's rocky shores on a calm January 1.

Opposite: The carcass of a tyrannosaur dinosaur bobs lifelessly in the Cretaceous Salish Sea.

Opposite right: A beach pebble full of fossil clams.

Ray had been in contact with Brandon and had prepared an image of an upside-down tyrannosaurid corpse floating in a sea full of ammonites. True to the concept of the eternal coastline, the sea has always been the sea, but it has been inhabited by some very different beasts over time.

My friend Tom Grauman invited Ray and me to join him in Friday Harbor on San Juan Island to celebrate the New Year of 2015 by boating to Sucia Island. For Tom it was an annual expedition. For Ray and me, it was a chance to see the site of Washington's only dinosaur. It was a surprisingly calm and clear January 1 as we slipped out of Friday Harbor and motored north around the west coast of Orcas Island. Sucia is the northernmost of the American San Juan Islands, and from the air it looks like the letter "E" with a couple of extra crossbars. Geologists immediately recognize it as a giant fold in the rock

The heteromorph ammonite *Eubostrychoceras.*

known as a syncline. The Nanaimo Group is a geologic formation that has mudstone, sandstone, and conglomerate, and it is the latter two that make the prominent folded ridges that form the many bays of Sucia. The sandstone cliffs of Sucia had been the source for much of the building stone of Seattle's Pioneer Square and most of Seattle's cobblestones. The patio of my dad's house in Beaux Arts is made of these cobblestones, a fact that I did not realize until this trip. The westernmost bay is called Fossil Bay, and that, obviously, was our destination.

A few hours later, we motored into Fossil Bay and tied up at the dock. Bull kelp floated in the bay and the sun warmly glanced off the red trunks of the madrone trees. The state park had recently installed a small interpretive panel about the fossils of the island, and we set off to see them for ourselves. Ray was really interested in finding more of the dinosaur, which seemed like a long shot to me – but you never know. I was more interested in seeing how many different kinds of ammonites we could see in the cliffs. The tide was high, but we were able to work our way to the western shore of the island and get onto the beach where the receding tide gave us more and more room to stroll.

The sun was out, the wind had dropped, and the surface of the sea was a glassy calm. It was one of those perfect winter days when the weather stands back and lets you enjoy life. We strolled along the rocky beach, looking both at the crumbling cliffs and at the beach cobbles. The cobbles were a mixture of granite pieces that had been delivered to the island by the recently departed ice sheet and chunks derived from the nearby cliffs.

Within minutes, we were finding a variety of clams including *Inoceramus*, the classic ribbed oyster of the Cretaceous, but also the fan-shaped *Pinna* and the elaborately flared *Trigonia*. Normally clams are far more common than ammonites, as they are bottom dwellers and ammonites are swimmers. Clams live where they are fossilized, whereas ammonites swim through the water column and sink to the seafloor when they die.

It was a perfect day, and we did not need too much patience before we found our first ammonite, a smooth-shelled *Gaudryceras*. Eventually we also found other ammonites, including a ribbed *Canadoceras* and the bizarrely coiled heteromorph ammonite known as *Nostoceras*. Our finds were adding up to the inescapable fact that the Salish Sea of 82 million years ago was very different indeed from the Salish Sea of today. We capped off the afternoon by finding a few coiled nautilus shells, which were further evidence of a different world. As expected, we found no evidence of dinosaurs, but it was comforting to know that we had been where Goedert had found Washington's first. We

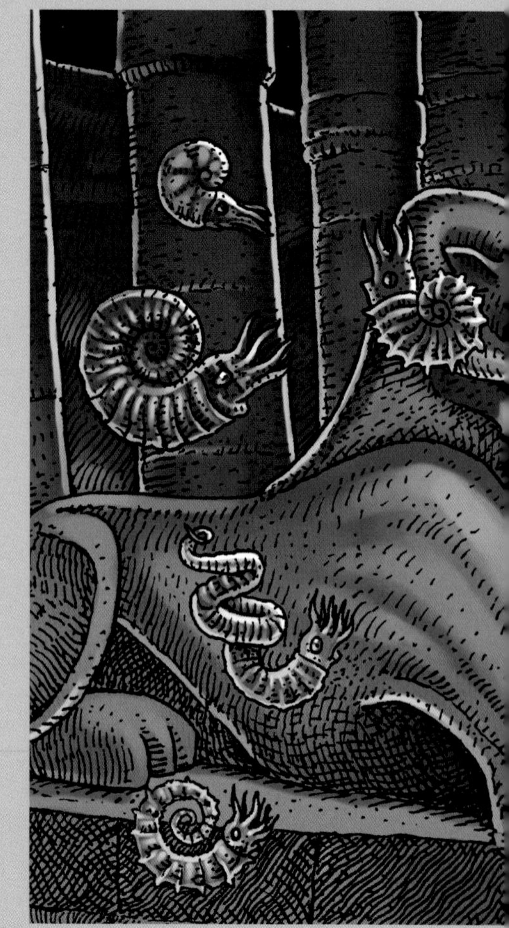

left the fossils as we found them, embedded in the cliffs, and headed back to Friday Harbor with the sublime feeling of a year well launched.

The trip to Sucia was the last part of our investigation of the northern Salish Sea. It was a few years earlier, on a sunny summer afternoon in 2012, that Ray and I landed in the Vancouver airport and rented a car. We had only given ourselves a few days to explore Vancouver Island, a task I knew should take months. Our main goal was to see the rocks of the Nanaimo Group and visit a group of amateur collectors who had been finding amazing fossils along the east coast of the 280-mile-long island. There is no bridge to Vancouver Island, so the only way to get there in a car is to take one of the many ferryboats that depart from Tsawwassen or Horseshoe Bay. As we sat in the ferry line, I started telling Ray what little I knew about the ammonites of Vancouver Island, and of all the places I had heard of but had never visited. As I spoke, I realized that we really didn't have a plan and maybe we needed one. When in doubt, call a paleontologist.

I called Peter Ward, a University of Washington professor and author, who had written his PhD thesis on the ammonites of the Nanaimo Group. Peter is a wildly creative paleontologist who has written more than a dozen well-received books about paleontology. Lately he has been studying the ammonites of Antarctica and the living *Nautilus* of the South Pacific. Back in the 1980s, he and I had been on the same side of the aster-

The Temple of Ammon

oid extinction argument. He found evidence that ammonites had suffered extinction precisely at the time of the asteroid impact, and I was seeing the same pattern with land plants. Like me, Peter had grown up in Seattle and had been strongly influenced by the Burke Museum. One time, he showed me a car-sized glacial erratic boulder in a park near downtown Seattle that was completely full of Cretaceous clam fossils. Peter answered my call on the first ring, and I told him that Ray and I wanted to be British Columbia ammonite tourists. To my amazement, he said that it would be easier for him to show us than to tell us and that he would catch a floatplane from Seattle the next morning and join us in Nanaimo. Problem solved.

We drove onto the ferry and crossed the Salish Sea, winding through the Gulf Islands that lay alongside the southeastern side of the island. The British Columbia Ferry system employs marine ecologists to give lectures to the passengers, and we enjoyed tales of squid, cormorants, and salmon, and we enjoyed the afternoon. As we approached Vancouver Island, I reminded Ray that the coast west of Victoria had outcrops that had yielded two different kinds of desmostylians: *Cornwallius* and *Behemotops*, and that we would soon be landing on the Saanich Peninsula that had once been home to the massive Ice Age *Bison latifrons*. We landed at Swartz Bay and drove into Victoria in time to visit the Royal British Columbia Museum.

This venerable museum opened in 1886 in response to the growing realization that people were coming from around the world to collect art and artifacts from the Native peoples of British Columbia, and that the province had better start saving some of it themselves. Today, the museum has three galleries: Natural History, First Nations, and the curiously named Modern History.

Peter Ward and a portion of *Diplomoceras*, the giant paper clip ammonite.

Opposite: The heteromorph ammonite *Polyptychoceras vancouverense*.

Opposite right: The heteromorph ammonite *Nostoceras hornbyense*.

The Native peoples or First Nations of British Columbia represent twenty-seven major groups that speak a total of eight languages. The coastal groups include the Haida, Tsimshian, Gitxsan, Nisga'a, Haisla, Heiltsuk, Oweekeno, Kwakwaka'wakw, Nuu-chah-nulth, Nuxalk, and Coast Salish. These were the people who were living along the coast when the first Europeans arrived, and these were the people who made the monumental cedar houses, ceremonial poles, and enormous boats that impressed the explorers and launched dozens of collecting expeditions. Their items can be seen in museums around the world, and the Royal BC museum contains amazing collections, including many poles that were salvaged from villages that had been abandoned due to smallpox epidemics at the end of the 1800s.

All of these people had made their living along the coast, and their beautiful hand-carved objects are full of images and references to whales, fish, eagles, ravens, beavers, seals, frogs, sharks, octopuses, bears, wolves, and other animals that shared the coastline. Both Ray and I have long been fascinated by these images and carvings and have revered the people that made them. The exhibit was packed with amazing artifacts, but I was really pulled to the stone carvings from Haida Gwaii (the archipelago formerly known as the Queen Charlotte Islands). Made from a silky black rock known as argillite, the carvings were like human-made fossils. The Natural History gallery contained real fossil ammonites from the ancient coastline, and the transition for me was seamless.

The museum also had a number of habitat dioramas that had been painted by Jan Vriesen, an artist who had worked with me in Denver. By chance, Jan was in town, and we met up with him for dinner at the home of Trish Guiguet, the daughter of the former, and longtime, curator of the Royal BC Museum, Charles Guiguet. Trish is a self-described "salmon person," and we dined in her backyard in the shade of a lovely *Metasequoia* tree. She told us tales from the 1940s of how the museum collected the animals and plants for the dioramas. Over dinner Jan decided to join us on our drive north. Now it was two artists to one scientist.

The next morning, we met the floatplane as it landed in Nanaimo. Peter Ward walked off the dock and leveled the score at two artists and two scientists. This was starting to get interesting. We drove past a pub called the Cranberry Arms, and I recalled that it was known to paleobotanists because a Cretaceous fossil leaf site had been discovered there in the 1990s when roadwork opened a road cut. The Nanaimo Group has rock layers that both formed as seafloor sediment and on land. This is a great combination, since the same sequence of rocks shows what life was like both onshore and at sea. The fossil plants included palms, cycads, broadleaf trees, and conifers. One particularly odd fossil plant from this site was called *Protophyllocladus.* Its closest living relative is a conifer called *Phyllocladus* that grows today only on the island of Tasmania off the coast of Australia. That was not an expected discovery.

It was the first time Ray and I had traveled with a personal ammonite specialist, and the conversation rapidly spiraled down the tunnel of cephalopod obsession. Peter connected the dots of the West Coast ammonite story for us. There are four regions along the West Coast that have yielded abundant Cretaceous ammonites: San Diego–Baja California, northern California, southern Vancouver Island (and Sucia Island), and the Matanuska Valley north of Anchorage. He had been working to figure out what the areas had in common and how they were different, and in so doing, reconstruct the world of the Pacific Coast of 80 million years ago. Many of the ammonite species could be found all the way from Baja to Anchorage, suggesting that the difference in climate from Mexico to Alaska was less in the past than it is today. One big difference was that the fossil sites in Baja preserved a type of fossil clam known as a rudist. These were big clams the size and shape of wastebaskets that formed tropical reefs in a world where the waters of the equatorial regions were too warm for coral reefs.

Peter also told us how he had been capturing, tagging, and releasing living nautiluses in the South Pacific to understand how they functioned and fed. He was really worried that the living nautilus was endangered because people loved their shells so much that they were driving them toward extinction. It was a bit ironic that ammonites had met their demise at the hands of the dinosaur-killing asteroid, and we were off to find ammonites with a nautilus expert who was opining about their pending extinction at the hands of humankind. It did make me feel better about collecting the already-extinct ammonites as opposed to the soon-to-be extinct nautilus.

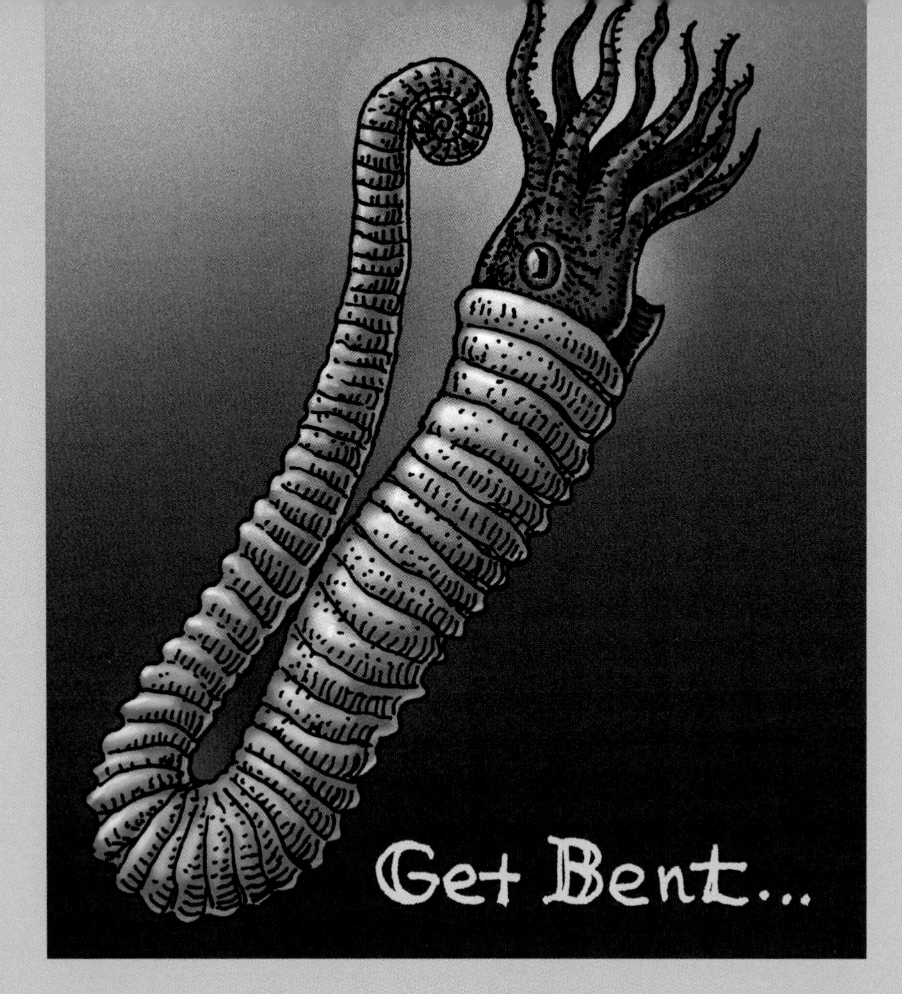


We headed to Denman and Hornby, a pair of islands off the east coast of Vancouver about 100 miles north of Victoria. Hornby is famous for producing some of Vancouver's finest ammonites, and the competition to find them has been intense. During the past fifty years, private collectors have scoured the island, and there are many fine collections in garages on Vancouver Island. The ammonites are found in concretions, and people have even taken to scuba diving to recover the precious nodules. I had long heard about the famous spot and really wanted to visit. Peter warned us that it required a double ferry ride to get there: one ferry to Denman and a second from Denman to Hornby.

We made the boat to Denman, and Peter took us to a site near the ferry dock. The rocks looked promising, so we spread across the rocky shoreline and began searching for fossils. After a few hours, we had found nothing. Even worse, we realized that we had missed the second boat to Hornby and our visit to this fossil Shangri-La would have to wait for a future trip. It became clear that our best chance of seeing Hornby fossils would be in a local garage.

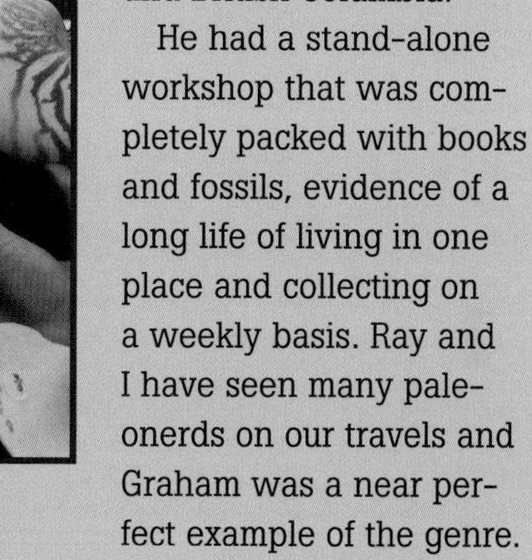

Peter Ward and a school of *Nautilus*.

We took the ferry back and headed to Qualicum Bay, a sleepy little town about 10 miles down the coast. We had come here to meet a local fossil enthusiast named Graham Beard. It was a sunny late afternoon when the two artists and two paleontologists pulled up in front of a modest bungalow. We were greeted by Graham and his wife, Tina. Graham was a short, silver-haired fellow in a striped shirt and khakis, and right away, I could see that he had the fossil disease. His yard was full of rocks and bones, and he got on topic immediately.

Graham was born in Wales and came with his family to Vancouver Island in 1947 when he was six years old. By the time he was sixteen, he had discovered the ammonites of Hornby, and his life was forever altered. He got the fossil bug bad, became a high school biology teacher, and devoted his life to the fossils of Vancouver Island and British Columbia.

He had a stand-alone workshop that was completely packed with books and fossils, evidence of a long life of living in one place and collecting on a weekly basis. Ray and I have seen many paleonerds on our travels and Graham was a near perfect example of the genre. Many of them have turned their basements, garages, or outbuildings into private museums. Graham had certainly done that, but he had also taken it a step further.

He invited us to take a stroll to the nearby Qualicum Beach Museum. Somehow he had pulled together the local support to enhance a small regional history museum with a very large fossil collection. The museum's tagline is "You'll find more than just a little history." It should be, "You'll find a ton of paleontology."

Graham welcomed us into the building like the proud founding father that he was. The small main room was filled with cabinets, cases, and incredible fossils. Graham had been collecting from Hornby for forty-five years, and the results were remarkable. Between his work and that of other amateurs and academic paleontologists, the Cretaceous marine outcrops of eastern Vancouver Island have yielded an amazing list of Cretaceous marine species. The list is a Cretaceous seaway "Who's Who" that includes ammonites, clams, snails, lobsters, crabs, pufferfish, ratfish, giant squid, frilled sharks, six-gilled sharks, seven-gilled sharks, bramble sharks, dogfish, saw sharks, nurse sharks, sand tiger sharks, horn sharks, sea turtles, the Cretaceous swordfish *Protosphyraena,* and even parts of the orca-sized marine reptile known as a mosasaur.

As we worked our way, fossil by fossil, around the crowded treasure room, we finally arrived at the reason I had wanted to come to Qualicum Beach. In 1979, one of Graham's high school students named Rosie had brought him a portion of a tusked marine mammal skull that her father had found while collecting oysters on a boulder-covered beach 6 miles north of Qualicum Beach. Graham contacted Rosie's dad, Bill Waterhouse, who, fortunately, had marked the spot. They relocated it and cleared the boulders away, exposing a slick, bluish clay surface. More bones were visible, and they started to dig. The fossil turned out to be a walrus that had been buried on its back with its head facing upward. In fact, it had been the vertically projecting tusks that had caught Waterhouse's eye in the first place.

Graham Beard and a cast of the skull of Rosie the Walrus

The clay was dense and sticky and dated back to about 70,000 years ago, the middle of the most recent Ice Age. The digging was difficult, and muddy water kept pouring into the pit and making it really tough to see what was going on. Graham recruited his ham radio friend Don McAlister to help with the dig. Don was blind, so he didn't care that he couldn't see what he was digging, and soon nearly 80 percent of the skeleton emerged from the Ice Age clay. Graham laid out the bones in his garage, and the fossil walrus found by an oysterman and excavated by a blind ham radio operator turned out to be one of the most complete fossil walrus skeletons ever found on the West Coast. In 1992, Canada's premier Ice Age mammal specialist, Dick Harington, described the skeleton and counted the growth rings on the tusks. He learned that it was a twelve-year-old female walrus. I was really excited to see the skeleton and was quite disappointed to learn that Graham only had a replica of the skull. The original skeleton had been recalled by Harington to the Canada Museum of Nature in Ottawa. We photographed Graham with the cast and cursed our timing.

It took me three years before I found myself in the offsite storage facility of the Ottawa museum and located the warehouse shelf that housed the incredible skeleton of what Graham had come to call "Rosie the Walrus."

By this time, we had been looking at fossils with Graham for several hours, and hunger was beginning to set in. We said our good-byes and went downtown to check into the hotel and get some food. Ray struck up a conversation with the hotel clerk and asked her if she was aware of Qualicum's walrus and its legacy of ammonites. She replied that she didn't pay too much attention to "fancy rocks."

The next morning, we headed north to Courtney to see another little museum that had been taken

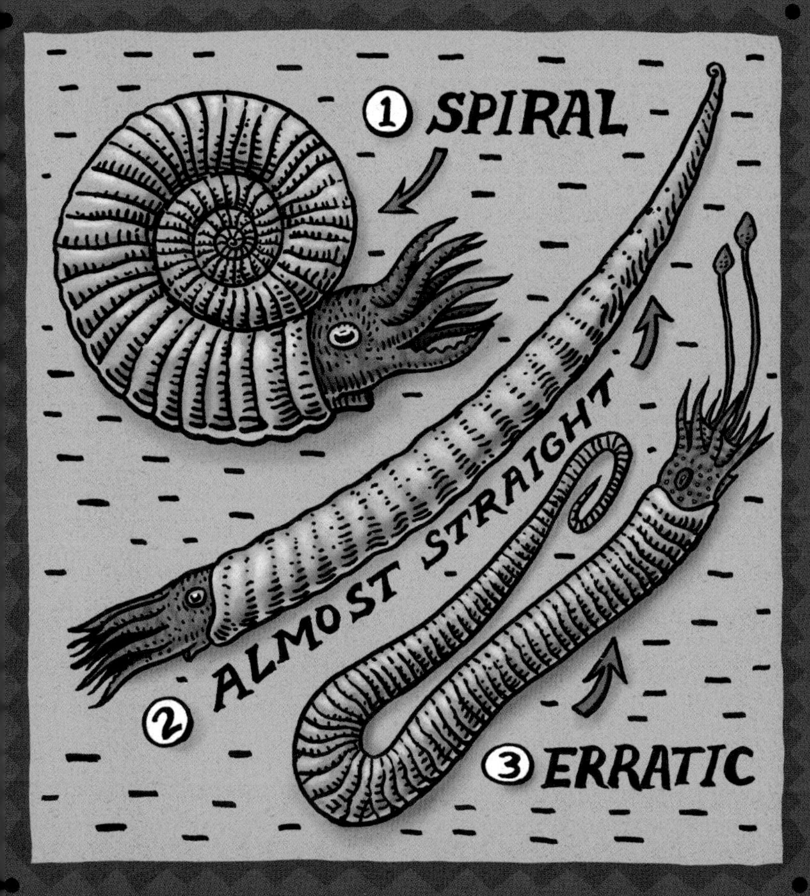

## The Three Shapes of Ammonites



If you pay attention to ammonites, you eventually realize that they come in three different shape categories. Most are some form of a spiral with variations based on how tight the coils are and whether or not the outer coils envelop the inner coils. Then there are straight ammonites known as the baculites, which look like oval rods that can be up to 5 or 6 feet long. The third type includes shells whose shapes appear to be erratic. These are the heteromorph ammonites, and they come in a host of configurations. The Nanaimo Group is known for its heteromorphs, and Graham showed us several. Peter was a heteromorph specialist, so he took over the conversation and told us about the mother of all heteromorphs, an ammonite called *Diplomoceras*. This was a big animal that had a shell that looked like an enormous 6-foot paper clip. Peter had found his first one in an outcrop farther up the coast. Over the years, he paid attention and eventually found examples on Seymour Island in Antarctica. Others turned up in Alaska, and Peter realized that this species had pulled off a very rare trick in biology. It is really unusual for one species to inhabit a range that includes both polar regions. Today, it happens with Arctic terns, orcas, and humpback whales. Apparently, the giant paper clip ammonite was the Arctic tern of the Mesozoic.

over by fossils. We were greeted by director Deborah Griffiths, curator Pat Trask, and several staff who were alerted that a team of paleontologists and artists were incoming. Pat told us the story of how this local history museum, which used to be all about "wagon wheels and arrowheads," had its transformational moment.

On November 12, 1988, Mike Trask, Pat's twin brother, and Mike's twelve-year-old daughter Heather, were hunting for ammonites on the banks of the Puntledge River. Heather found a string of vertebrae exposed in the outcrop, and it was immediately clear that she had found one of the larger denizens of the Cretaceous Salish Sea. Mike and Heather collected a few bones and sought the advice of Betsy Nicholls, the marine reptile curator at the Royal Tyrrell Museum in Drumheller, Alberta. Betsy confirmed that the bones belonged to a long-necked plesiosaur. This discovery created a lot of local excitement; no one had ever found such a thing on Vancouver Island. The museums in Vancouver and Victoria didn't have the paleontologists to take on a full excavation, so the local amateur paleontologists organized themselves and started to extract the skeleton.

The excavation proceeded, and much to everyone's amazement the skeleton was pretty complete and nearly 40 feet long. Courtney now had its own Loch Ness Monster, but this one was real. Betsy Nicholls stepped in and wrote letters to the provincial government arguing that Courtney should be allowed to collect and display the find.

The Puntledge elasmosaur proved to be a magnet for kids. It wasn't a dinosaur, but it was darn close, and it proved to be the catalyst for the revitalization of the Courtney Museum and for the rapid growth of an amateur group known as the Vancouver Island Paleontological Society. It turned out that there were lots of local collectors who had been finding fossils and stashing them in their basements. Now they had a place to bring their finds, and their finds attracted professional paleontologists who wrote books and papers and validated the importance of their discoveries.

We lingered in the museum for a couple of hours, chatting with the staff and looking at the fossils that had accumulated in the collections room in the basement. Not only were they housing fossils from Vancouver Island, but they were also receiving donations from around the province. With only a few professional paleontologists

### That Distinctive Spiral Shape

**Just a few months after we visited Graham, he got call from a woman who was hiking with her son near Benson Creek Falls just west of Nanaimo. The woman had stepped over a tire-sized depression in the bedrock of a dry creek bed without noticing anything, but her son noticed that the depression had a distinctive spiral shape. Graham visited the site and confirmed that the depression was the mold of a giant ammonite. Unfortunately, the actual fossil had already washed down the creek, but he was able to make a fiberglass cast that preserved the shape of the massive shell.**

Rosie the Walrus (*Odobenus*) from Qualicum in the vaults of the Canada Museum of Nature in Ottawa.

and remote nature of British Columbia means that its fossil treasures largely wait to be discovered.

Betsy Nicholls herself had worked on remote British Columbian fossils. In 1992, she was alerted to a large bone exposed in a river in isolated northern British Columbia. It took her and Makoto Manabe, my graduate school buddy from Tokyo and Japan's National Science Museum, several years to excavate what turned out to be a massive Triassic ichthyosaur that they named *Shonisaurus sikanniensis*. At nearly 70 feet long, it was the size of a big whale and the world's largest ichthyosaur.

Most of the Pacific Coast north of Vancouver is not accessible by road. This becomes immediately clear if you have a window seat on the flight from Seattle to Ketchikan. The entirety of the flight is a panorama of jagged snow-capped peaks and deep green fjords. From the point where Highway 99 departs Vancouver headed toward the northeast to the spot where Highway 37 arrives in Prince Rupert, there are literally no roads that access the British Columbia coastline for 450 miles. This huge and inaccessible coast is known as the Great Bear Rainforest, and it is home to a very unusual subspecies of black bear that is totally white. Known as Kermode bear or the spirit bear, it is one of the least-known animals on the West Coast.

working in British Columbia and at the museums in Vancouver and Victoria, Courtney had become the beating fossil heart of the province.

Nothing demonstrated this more than a massive ammonite in the back room of the Courtney Museum. A mining company had found the fossil near Fernie in southeast British Columbia and had searched for a museum to house it. They found Courtney. Ray and I had been hearing about a massive ammonite in the hills near Fernie, and we had hoped our travels would take us there, so we were stunned to realize that the giant ammonite had found us. We crawled over the 5-ton boulder and inspected the 6-foot Jurassic ammonite, realizing once again that the huge size

The people that travel this coast do so by fishing boats, tugboats, ferries, coastal freighters, private yachts, and cruise ships. In the winter of 1984, three of my friends and I found out that for only $75 Canadian, it was possible to buy a round-trip ticket with a four-person berth on a British Columbia ferry from Vancouver to Prince Rupert and then across to Queen Charlotte City on Haida Gwaii. Haida Gwaii is a remote archipelago located 50 miles off the coast of British Columbia. Home to the Haida people, it is known for its remarkable and monumental cedar sculpture, houses, and poles. I had seen historic photographs of dozens of totem poles in the village of Skidegate and more recent photographs of moss-covered totem poles at the abandoned village of Ninstints. The remarkable new building at the Museum of Anthropology at the University of British Columbia had opened in 1976 and had startled the museum world by opening all of its collections for public display. These collections contained impossibly beautiful argillite carvings from Haida Gwaii. I had also read literature from the 1800s that showed images of a diversity of Jurassic ammonites from the island. It really seemed like these remote islands were a simple way to travel back in time. I had to see this place. The price seemed too good to be true, and we couldn't help ourselves — we booked four tickets. We left Vancouver in late December in pouring rain, and it rained steadily for the entire run to Prince Rupert. My photographs from this trip show dark, wet forests and choppy green water. In Prince Rupert, we boarded another ferry and were rolled and battered by heavy ocean swells in the overnight crossing.

We had made no plans but were fortunate to rent a pickup truck and a hotel room, and we spent the next few days exploring the north island from Skidegate to Massett. We saw literally hundreds of bald eagles and deer. The resurgence of Haida art that was led by Native artists Bill Reid and Robert Davidson in the 1970s was under way, but we were disappointed to see that only one pole was standing in Skidegate. Then we learned it was impossible to get to Ninstints in the winter. It poured rain in a way that I had never experienced in Seattle, and sunrise was late and sunset was early so our days were sodden five-hour experiences. On the last day, the skies cleared and we rented a salmon trawler from a fisherman who knew of a small island full of fossils. The wind ceased and the sound was glassy calm. Huge Sitka spruce cast their reflections on the coils of bull kelp floating in the water. The island's beaches were full of concretions and ammonites. The place was remarkable.

I didn't know it at the time, but marine biologist Ed Ricketts, who we met in Chapter 4, collected intertidal animals on Vancouver Island in 1945 and 1946 and in Haida Gwaii in 1946. After his trip to the Sea of Cortez with Steinbeck in 1940, he had conceived a series of West Coast collecting trips that would take him as far north as the Bering Sea and allow him to describe the intertidal ecology of the entire West Coast. On Vancouver Island, he spent his efforts on the west coast near Tofino, and in Haida Gwaii he collected along the north coast near Masset. These travels, and Rickett's relationship with John Steinbeck and also mythologist Joseph Campbell, are beautifully described in Eric Tamm's book, *Beyond the Outer Shores*. Ricketts considered Haida Gwaii, then known as the Queen Charlotte Islands, to be the Galapagos of the north, and we can only wonder what would have happened to our view of this remote place if he had lived to finish his book about them. Ricketts was only weeks away from departing for Haida Gwaii with John Steinbeck in 1948 when he was hit and killed by the train on Cannery Row.

The book that did get written about Haida Gwaii is John Vaillant's *The Golden Spruce*. This book tells the story of Haida people, of the endless forests of the Great Bear Rainforest, of European contact, and of the logging that has decimated the forests of Canada's coast.

Both Vancouver Island and Haida Gwaii are part of a landmass known as Wrangellia that assembled somewhere out in the Pacific before sliding into place and docking up against North America about 200 million years ago. These exotic islands are exotic for a reason. They are not of this continent. But after they arrived, they became part of the West Coast and part of its story. Over time, the plants, animals, and people of North America (many of them originally from Asia) populated the islands and created ecosystems and cultures. The Haida creation story, so eloquently rendered in Bill Reid's giant yellow cedar sculpture, tells of the raven discovering the first humans living in clamshells along the beaches of Haida Gwaii.

Back on the mainland, in the southeastern corner of British Columbia, is a fossil site known as the Burgess Shale. It is the most famous example of the Cambrian Explosion, a time more than 500 million years ago when many of the world's first marine organisms (including the ancestors of clams) appeared on earth. That is a place that tells the paleontologist's creation story.

Opposite: The Puntledge plesiosaur in the Courtney Museum.

Left: Jurassic ammonite from Haida Gwaii.

Bison priscus
Bison
Ring Eider
Saurornitholestus
Lambeosaur from the Colville River
Ammonites from Northern Yukon
Steppe Bison
Bowhead whale
Barrow
Wainwright
From Prince Creek Formation
Paleo-Walrus
Ginkgo
Pachyrhinosaurus perotorum
Nanuqsaurus hoglundi
Steppe Mammoth
Fox
CHUKCHI SEA
Vankarem
Saiga Antelope
Mastodon
Mesozoic Plant Fossils
Cape Lisburne
Point Hope
Polar Bear
Prudhoe Bay
Deadhorse
Kaktovik
Hyena from Old Crow
Pleistocene Walrus
Ugrunaaluk kuukpikensis
National Petroleum Reserve
Triassic Ichthyosaur
Troodon
Colville River
Shublik Mountains
Arctic National Wildlife Refuge
Ivvavik National Park
DE LONG MTNS.
Kivalina
Campylognathoides
Woolly Rhinoceros
Devonian Lungfish Dipterus
BROOKS RANGE
Alaskacephale
BAIRD MOUNTAINS
Buzz Saw Shark!
Lycoceras
Helicoprion
Homotherium
Ice Age Wolf
RUSSIA
Noatak
Kobuk Valley National Monument
Anaktuvuk Pass
Gates of the Arctic Nat Park
Triassic Coelacanth
The Vanibou
PHILIP SMITH MTS.
Jefferson's Ground Sloth
Beringian Lion
Kotzebue
Canis lupus
Pacific Walrus
Helmeted Musk Ox
Steppe Bison Bison priscus
Shishmaref
Paleo Pika
SCHWATKA MTS.
Homotherium
Wiseman
Coldfoot
Whitestone Mammoth
Wales
Provideniya
Port Clarence
King Island
Teller
Bering Land Bridge National Preserve
Short Faced Bear Arctodus simus
Elephant Point
Yak
Alaska Pony
Camelops
Fort Yukon
Steppe Bison
Paleogene Sequoia Fossils
Gambell
Savoonga
St. Lawrence Island
Musk Ox
Nome
Equus
Bootherium helmeted musk ox
Stag Moose Cervalces scotti
Pleistocene Horse
Beringian Lion
Yukon River
Steppe Bison
Woolly Mammoth
Steller's Sea Cow until 1900 AD
Beringian Lion Skull
Young Woolly Mammoth
Shaktoolik
Unalakleet
Galena
YUKON
Therizinosaur footprints
Short Faced Bear
Fairbanks
"Blue Babe"
Eagle
Mammuthus primigenius
BERING SEA
Steller Sea Lion
Bering Wolffish
Castoroides giant beaver
Dino Tracks
Ornithopod Dino
DENALI NATIONAL PARK
Dinosaur Footprints
UA Museum of the North
Delta Junction
Chicken
Dawson City
Saiga Antelope
Giant Short Faced Bear
YUKON CANADA
Dragon Poacher
Walleye Pollock
Hooper Bay
Mastodon Tooth
Sauropod Tracks
McGrath
Talkeetna Mountains
Pachydiscus Ammonite
Crinoids
Fossil Weasel
Brontothere
ALASKA
Palaeotyphlops
Paleocene Bird
Mastodon
Mammoth
Nunivak Island
Bethel
Anchorage Museum
Conodonts
Anchorage
Wasilla
Palmer
Valdez
Copper Center
Kennecott
Paleogene Palms, Ferns and Leaves
Wrangell St. Elias National Park
Dall Sheep
Mountain Goats
Red King Crab
Steppe Bison
YUKON DELTA NATIONAL WILDLIFE REFUGE
Giant Bird
Cretaceous Fossils
Mount Redoubt
Cook Inlet
Seward
Prince William Sound
Alaska Sea Life Center
Cape Yakataga
Cape St. Elias
POLAR BEAR AND MAMMOTH COMBO
St. Paul
PRIBILOF ISLANDS
Orcinus orca
Atka Mackerel
Gray Wolf Canis lupus
Platinum
Dillingham
Steller's Sea Eagle
Iliamna
Kokhanok
Augustine Volcano
Homer
Diplomoceras Ammonite from Matanuska Valley
Silurian Brachiopods
St. George
Pacific Halibut
Pacific Spiny Lumpsucker
Round Island
BRISTOL BAY
King Salmon
KATMAI NAT. PARK
Oligocene Blackfish
Salmon Shark
Oligocene fossilized Ratfish
Silurian Gastropods
Ounalashkastylus tomidai
Pacific Walrus Odobenus rosmarus
Canadoceras yokoyamai
Pliosaur
Megalneusaurus
GULF OF ALASKA
Kodiak
KODIAK ISLAND
Didymoceras
Humpback Whale Megaptera novaeangliae
Cheirurus trilobite
Desmostylian
Kolponomos
Desmatophoca
Pachydiscus kamishakensis
Kammerkarites
Cadoceras
Desmostylian from Kodiak
CHIRIKOF ISLAND
"There be monsters here..."
Sperm Whale Physeter macrocephalus
Pleistocene Ringed Seal and Caribou from Shuká Kaa Cave Prince of Wales Island
Shonisaur
Unimak Island
Unalaska Island
Umnak Island
Nikolski
Dutch Harbor
SHUMAGIN ISLANDS
Petrified Forest
Inoceramus Cretaceous Clam
Sleeper Shark - Somniosus pacificus
Steller Sea Lion

# ALASKA AND THE YUKON

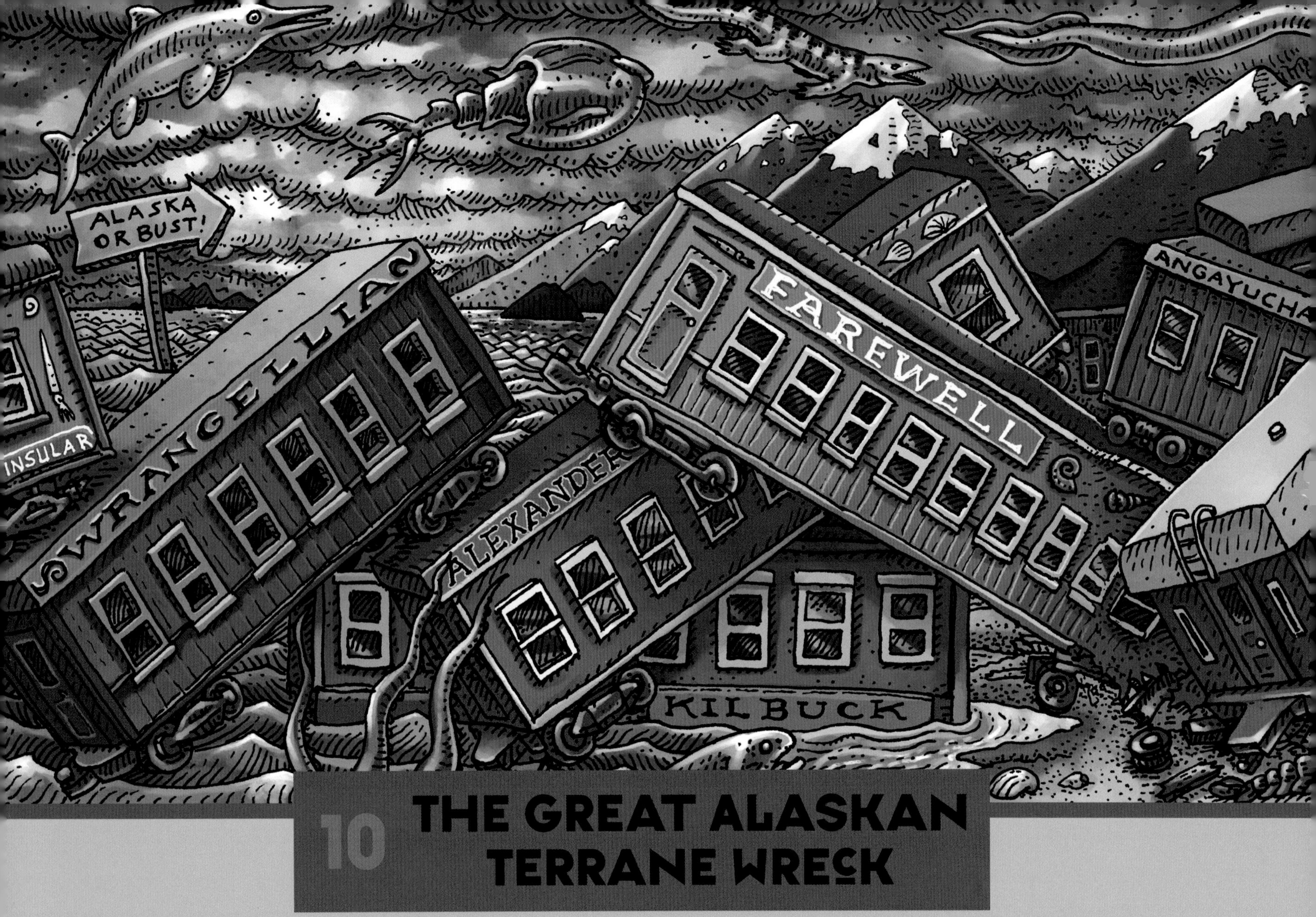


# 10 THE GREAT ALASKAN TERRANE WRECK

Once you get used to the idea that continents move, it's not hard to imagine that little pieces of continents can be shoved around like the spare parts that they are. John McPhee writes about this in his book *Assembling California*, and by the early 1980s, geologists were beginning to become aware that most of the west coast of British Columbia and Alaska had this type of very complicated history.

They called these little continental fragments "terranes" and talked about "exotic terranes" and "suspect terranes." Suddenly the complex geology of southeastern Alaska had a conceptual framework that started to make some sense. Today, some geologists call this the "Great Alaskan Terrane Wreck."

While I was building my interest in Alaska and working in California, Ray simply moved there.

The exotic geologic terranes of coastal Alaska depicted as train cars crashing into the western edge of North America.

He had been living in Seattle when his sister started a fresh fish stand in Ketchikan and invited him up to help out for the summer of 1983. He found that he liked the small town at the southern end of the Alaskan panhandle and decided to stick around. Up until that time he had been an artist, musician, and photographer. Once he got to Ketchikan, fish started to become an increasingly important part of his life.

It's not hard to see why. Ketchikan is a tiny town on an island whose main industries were fishing and logging, and the area still has plenty of fish and trees. Unlike Puget Sound, which had been largely logged and fished out, southeast Alaska was searching for ways to make the harvests sustainable. In the summers, Ketchikan would fill with seasonal cannery workers and tourists. In the winter, it would quiet way down and the rains would come. Some people think that it rains a lot in Seattle, but Seattle's 38 inches are nothing compared to Ketchikan's 156 inches.

Once in Ketchikan, Ray started drawing fish and he never looked back. In 1984, he had the idea of silk screening one of his fish images on a T-shirt and found that they sold easily. For a guy who had spent his formative years in Kansas, Alaska was a wonderful new world. He reveled in the majestic Tongass rain forest, the incredible diversity of the sea, the proximity to Tlingit and Haida artists, and the small-town scene. Fish became his muse and he became their artist. His T-shirts sold well and paid the bills, but his quixotic focus on the real details of the natural world soon caught the eye of scientists who saw in Ray an opportunity to promote the organisms they studied and loved. I can remember buying Troll T-shirts in Seattle in the early 1980s and wondering who was doing such interesting work.

As Ray settled into his life in Ketchikan, he began to meet Alaskan scientists who introduced him to some of the more remote and subtle opportunities of the landscape. He found that he could scratch the itch of his childhood love of fossils by visiting particular beaches where fossils could be found.

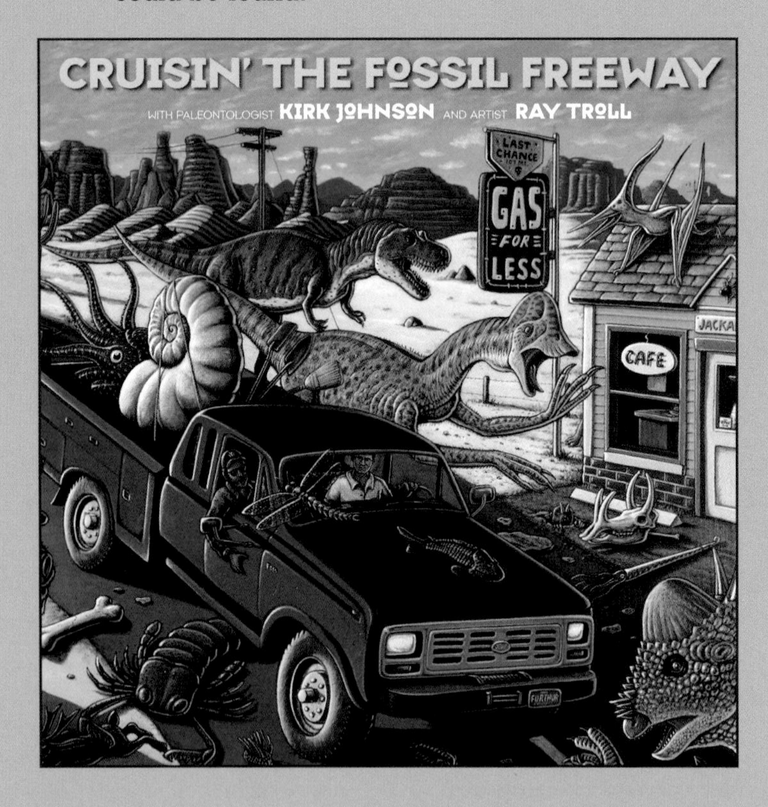


I finally met Ray in 1993 and made a point to visit him in Ketchikan a few years later. By 1998, we began to work together and spent almost nine years researching and writing the book *Cruisin' the Fossil Freeway: An Epoch Tale of a Scientist and an Artist on the Ultimate 5,000-Mile Paleo Road Trip.* I was living in Denver at the time, and that book focused on the classic fossil fields of the American West. Our 5,000-mile road trip included Colorado and Wyoming and parts of all of the other states that surrounded those two. Even as we were working on that book, we knew that our interests would ultimately take us to Alaska.

The inside passage from Seattle to Juneau is most frequently traveled by the cruise ships that now arrive, several per day, in the towns of southeast Alaska. Amazingly, nearly a million cruise passengers will visit Ketchikan each summer.

For me, going to Ketchikan is like going back in time in Seattle. It is a place of rain, big fish, big trees, local and Native artists, and unexplored opportunity. Visiting Ray means dropping by the carving shed at Saxman Village where master Tlingit craftsmen like Nathan Jackson and Haida carvers like Donny Varnell carve huge totem poles, or heading down to the salmon hatchery to go snorkeling with schools of spawning salmon (remember to watch out for bears while snorkeling).

Over the years, Ray had gotten to know local geologists and had begun to learn about the complex geology that was all around him. He hit the jackpot in 1991 when he met Jim Baichtal, a remarkable example of that species known as the Alaskan geologist. Jim is the forest geologist for the Tongass National Forest and is based in Thorne Bay on Prince of Wales Island, where he is responsible for thousands of square miles of extraordinarily rough and inaccessible terrain. From a fairly close distance, Jim looks like a nineteenth-century mountain man with his extravagant silver beard and logger's suspenders. If you search for him online, you will see him posing with immense moose, deer, mountain sheep, and musk oxen, all killed with his black powder musket. He moves easily around his world, traveling by foot, canoe, pickup truck, landing craft, float-plane, and helicopter. Jim is quick with a laugh, and he is a font of detailed local information. He has made amazing discoveries in this complex landscape.

Jim Baichtal, the geologist of the Great Alaskan Terrane Wreck.

If you look at a map of southeast Alaska, you will see a dense potpourri of islands and channels. And if you're like me, you'll ask why it looks that way. Again, the answer is complicated. The origin of the massive Coast Ranges of British Columbia and Alaska has vexed geologists for more than a hundred years, but the pattern is starting to come into focus. Many of the channels represent giant geologic faults. A fault is a crack in the Earth where one side has moved relative to the other. Once faults form, they are often places of enhanced erosion and thus often become the low points in the topography. It now appears that the many terranes of southeastern Alaska have been shoved into place over time, and the boundaries between these terranes become the channels. The bedrock here includes rocks that formed first as sediments, rocks that erupted as lavas, rocks that began as reefs, and rocks that have been tortured and changed by intense heat and pressure. All of these rocks are covered either by the saltwater of the channels or the temperate rain forest and its shallow but luxurious soil.

This means that the best place to see the rocks is right at the shoreline where the plants can't grow. This part of Alaska has large tides, so timing is everything and many good sites are only exposed at low tide.

This faulted, forested geology is further complicated by the fact that it too was recently glaciated, and many of the islands were covered by as much as 3,000 feet of ice as recently as 10,000 years ago. When ice covers a landscape, its weight pushes the land down. When the ice melts away, the land rebounds up. This process of landscape rebound (sometimes called glacial or isostatic rebound) can happen pretty fast. There are places in Scandinavia where old Viking docks are more than a hundred feet above sea level. Remember that when ice melts the sea

## Of Ancient Sharks and Cheeseburgers

When I was working on the *Planet Ocean* book with Brad Matsen in 1993, we toured the back rooms of the Natural History Museum of Los Angeles County with a paleontologist named J. D. Stewart. He pointed out a curious fossil lying on the floor. It was a perfect spiral of shark teeth from a critter called *Helicoprion*, and it became my personal obsession. Soon, I was contacting experts from around the world and reading arcane scientific literature. It turned out that there was a whole menagerie of these underappreciated ancient fish.

Jim Baichtal and I came across a geologic report that mentioned shark teeth had been found in Devonian rocks on the shore of Suemez Island, off the outer coast off Prince of Wales Island. In the spring of 2004, my buddy Tom Fowler and I chartered a floatplane to Craig, Alaska, where we met up with Rich Manning. Rich was friend of Tom's who owned a small fishing lodge named Catch a King with a fleet of well-equipped seaworthy boats. The fishing season hadn't gotten under way yet, and Rich had agreed to take us to Suemez for the day. Jim met us there at the boat dock that morning, along with his trusty rock hammer, Thor.

Our "treasure map" for the expedition was a bad photocopy of a bad photocopy from 1975 with a couple of small X's marked on an island in a bay just off Suemez Island. We walked the coastline of the island all morning and into the late afternoon and only had a handful of Devonian brachiopods to show for it. We decided to go out and fish the tide change near Noyes and enjoy some petroglyphs. I was pretty bummed out and sat sullenly on the back deck of the boat. I pulled out the map again and stared at it for a while. Then I realized the small blurry X wasn't quite where we'd thought it was. We had spent the whole day searching on the wrong freakin' island!

As the afternoon shadows grew longer, we headed to the dock in Craig. I pleaded with the crew to turn around. After all, when would we ever have another chance? But my whimpering fell on deaf ears. Everyone was tired and hungry and ready to be back at the bar. Desperate, I offered to buy cheeseburgers at Ruth Anne's restaurant for everybody if they would give it one last try, and they agreed. We hit the beach at the new spot just as the sun was starting to go down and the no-see-ums were coming out in full biting force. Jim, Tom, Rich, and I spread out and walked the beach, kicking rocks aside and cracking a few along the way. Within a few minutes, Jim yelled, "Bingo!" We had our shark.

I don't know what it is about the two of us, but when we hunt fossils together, we have yet to get skunked. It really is a matter of the proper "search image." As soon as Jim noticed that the small black specks were actually double-pronged teeth, we began to see them all over the place. The outcrop of limestone was only a couple of feet wide and stretched for about 30 feet before it tapered off and disappeared below a layer of schist. As we got down on our hands and knees and looked closer, we began to notice larger blackish forms running through the rocks. We realized they were long banana-shaped shark spines. The cheeseburger bribe had worked.

After the trip, we sent a few of the smaller rocks back to John Maisey at the American Museum of Natural History in New York City. John's a widely noted expert on fossil fishes, and he soon identified the teeth and spines of several shark species. Looking back on it now, I realize that I'd achieved the "holy trinity" of fossil collecting. I had found my own Alaskan trilobite, ammonite, and shark tooth. Amen.

– Ray Troll

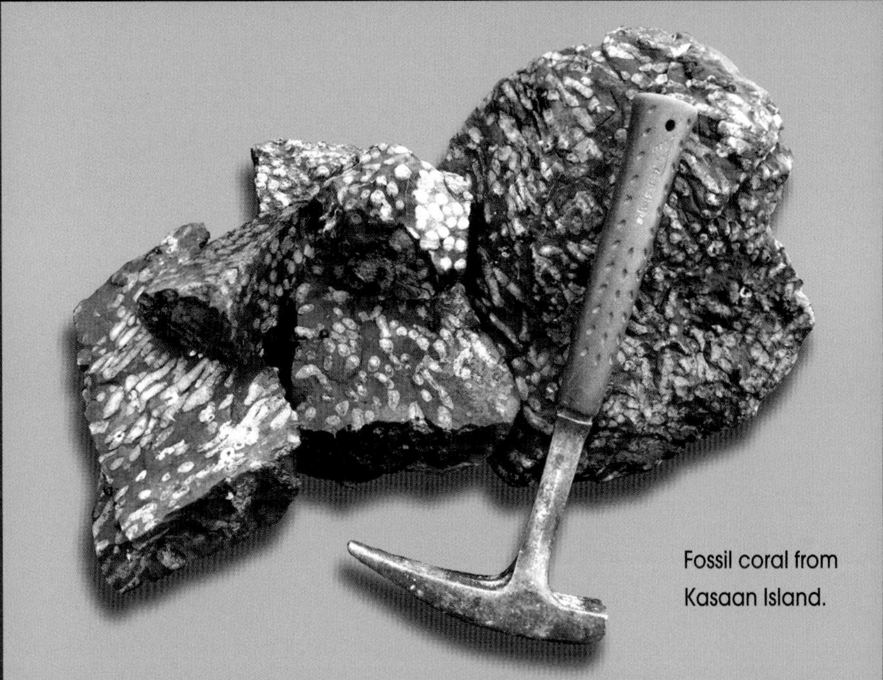

Fossil coral from Kasaan Island.

level goes up, so there are two processes happening in the same place. The land itself can go up and down and the sea level can go up and down. Welcome to Baichtal World. Jim is a master at understanding the evolution of the Alaskan coastline, and he uses it to locate archaeological sites with amazing reliability.

There is one final tool in his tool kit. The islands of southeastern Alaska are composed of parts of many exotic terranes, and many of these were first formed at lower latitudes in tropical climates. That means that there is a lot of ancient marine limestone mixed in with the other rock types in the area. Since limestone is formed of calcium carbonate, it dissolves in acidic conditions. The high rainfall in the Tongass Forest and the action of the forest soil creates these conditions, and many of the limestone deposits dissolve to form cave systems and jagged limestone surfaces. As the limestone dissolves, it often reveals fossils of the reef animals that made the limestone in the first place. This is known as a karst landscape and it explains why Jim is also a caver.

### Unlikely Neighbors

Caves are perfect places for bears to hibernate, which makes Alaska an exciting place to go caving and also makes caves a great place to find fossil bears. Remains of black bears were found in a newly discovered passage in El Capitan Cave on the north end of Prince of Wales Island in 1990, and, using the carbon-14 method, the bones were dated to show that the bears had lived between 11,500 and 12,300 years ago. These dates in turn showed that the ice sheet had retreated from the area by that time. This means that is it possible to use fossil bears to assess the presence or absence of ice sheets.

Things got even more interesting in 1996 when scientists excavating in On Your Knees Cave on Prince of Wales found the 9,730-year-old remains of a human. These are some of the oldest human remains in Alaska and show that people were sharing caves with bears in there as they were in Europe.

Ketchikan is located on Revillagigedo Island, and its airport is located across a narrow channel on Gravina Island. The airport is about the only thing on the island, so the proposal to build a $400-million bridge became the famous "Bridge to Nowhere" controversy of Sarah Palin fame in 2006. The west side of Gravina is composed of Triassic rocks. In 1993, Ray and Jim got together, pored over maps and old field reports, and began to explore these rocks for fossils. In 2013, they hit pay dirt and discovered bones from Triassic marine reptiles known as ichthyosaurs.

In 2006, they found the teeth of Paleozoic sharks on Suemez Island and lured John Maisey, one of the world's leading experts in fossil sharks to venture out from New York City to Ketchikan to get involved in the research. Armed with his relentless curiosity and an amphibious mountain man, Ray has set up a fossil franchise in one of the rainiest towns on the coast.

I wanted in on this fossil action, so Ray and I chartered a salmon fishing boat from Ray's friend Chip Porter, and we set off from Ketchikan toward Prince of Wales Island where we planned to visit a fossil site on the Alexander terrane. Our target was a place called Kasaan. If you Google Kasaan Village, you will see amazing photographs of an old Kaigani Haida village with remarkable totem poles and cedar houses. One of these houses, the Son-i-Hat Whale House, stands today. It was built around 1880 and restored in 1938. If you visit the Smithsonian's National Museum of Natural History, you will see three magnificent poles all collected in 1875 by James G. Swan for the 1876 Centennial Exposition in Philadelphia. These were the first three poles from the Pacific Northwest to be exhibited to the American public. One of the poles came from Haida Gwaii, the second from a village near Prince Rupert, and the third one from Kasaan. Swan had tried to buy a pole in Kasaan but the owners wouldn't sell. Instead they offered to carve him a replica for a fee. The Kasaan pole is the first instance of a commissioned pole, and I walk past it every day on the way to my office.

We were headed to Kasaan Island, which lay in the channel about 3 miles southeast of the village. Ray had been there with Jim Baichtal in 1991, and they had each found a trilobite. On the boat ride back to town, they discovered that they had each found part of the same trilobite. Trilobites are some of the most coveted of fossils,

and somehow, Ray talked Jim out of his piece.

It was a calm, cloudy day with the occasional drizzle. Chip was happy not to be fishing, and we were happy to be on an Alaska fossil trip. The tide was about halfway out when we pulled into a shallow bay. Chip launched a little rowboat to get us to shore, and we landed on the rocky, rubbly beach. Right away I started seeing fossil corals – the bedrock of the beach was a fossil coral reef. These fossils were Devonian, or about 420 million years old, and the reef had formed on a sliver of a continent that crashed into North America sometime later. Ever since, its lot was tied to North America.

Connie Soja is a paleontologist at Colgate University who studies fossil reefs of this age, and she has analyzed the fossils from Kasaan Island in an attempt to figure out where the Alexander terrane came from before it docked onto North America. Amazingly, the most similar reefs to Kasaan are found in the Ural Mountains of Russia. This means that the Alexander terrane came from a long way away.

Upper: The Haida village of Kasaan in 1908 (Photo courtesy of Alaska State Library, PS77-005-077 Kasaan )

Lower: Landing at Kasaan Island to find fossils.

Chip rowed ashore, and we built a little fire underneath a ledge made of fossil coral. Then we started crawling down the rubbly beach, three middle-aged guys, in the rain, in raingear, looking for fossils. The rain dampened the surface of the rocks, revealing the patterns within. As the rain continued, the beach boulders revealed themselves to be chunks of a long-gone coral reef. We found no trilobites. The rain continued to spit down and we crouched around the fire and thought about the incongruity of an ancient coral reef from Russia ending up on an Alaskan beach next to an old Haida village.

A few months later and at Connie's urging, Ray and I chartered a floatplane out of Juneau and flew 60 miles to the mouth of Glacier Bay. Here is a place where the retreat of glaciers is a remarkable process that you can literally watch. Early explorers who visited the bay in 1790 reported that a massive wall of ice extended to the very mouth of the bay. By the time John Muir visited in 1879, the retreating ice wall was 45 miles up the bay. As we flew along the bay, we could see that the ice face was now separated into a number of valley glaciers and was 65 miles from the mouth; a retreat of 65 miles in 225 years works out to about 500 yards a year. We landed in the bay and glided our floatplane to a smooth rocky surface that just a hundred years ago had been covered by ice. The glacially polished rock face exposed something I had never seen before: here in cross section was evidence for a 450-million-year-old landslide off the face of an ancient sponge reef. The reef had broken into

## Disappearing Glaciers

**In a warming world, glacial ice is the canary in the coal mine, and Glacier Bay shows us just how fast ice can retreat. The story is complicated, though. All along the Alaskan coast from Glacier Bay to Prince William Sound, glaciers are retreating up their valleys and leaving open water and bare rock in their place, but in a few places, they are advancing. As we flew back to Juneau, we saw the massive Juneau Icefield, which, at 1,500 square miles, is the fifth largest in the Western Hemisphere. Research on this ice sheet and the dozens of valley glaciers flowing out of it are helping scientists unravel the complexity of how glaciers advance and retreat.**

A glacier at the head of Glacier Bay.

large blocks and had tumbled down a submarine slope before lodging in the mud at the seafloor. The entire jumble had turned to rock and had traveled with the rest of the Alexander terrane before being uplifted and exposed at the surface. Then it was covered and polished by glacial ice, and finally the glacier had gracefully retreated to display the polished surface for our viewing pleasure.

One of the prime challenges of geology is that you have to be very nimble when thinking about time, space, and process. During our brief flight, we had seen evidence for a 450-million-year-old landslide that happened in a matter of minutes. The jumbled remains of that landslide were slowly carried across the entire Pacific Ocean at a rate of about 1 inch per year before slowly docking with North America. Then we had flown up a 45-mile section of open water that had been completely covered in glacial ice in 1879, the year my grandfather was born.

We landed in the bay and taxied up to the dock. Driving back to Juneau, we stopped at the Mendenhall Glacier, one of the many valley glaciers flowing off the Juneau Icefield. This glacier has retreated 1.75 miles since my mom was born in 1929. When you can measure geological phenomena in miles and family members, you realize that we really do live on a dynamic planet.

Upper: Landing on the shores of Glacier Bay.

Above: The square blocks on this glacially planed surface are cross sections of an ancient submarine landslide.

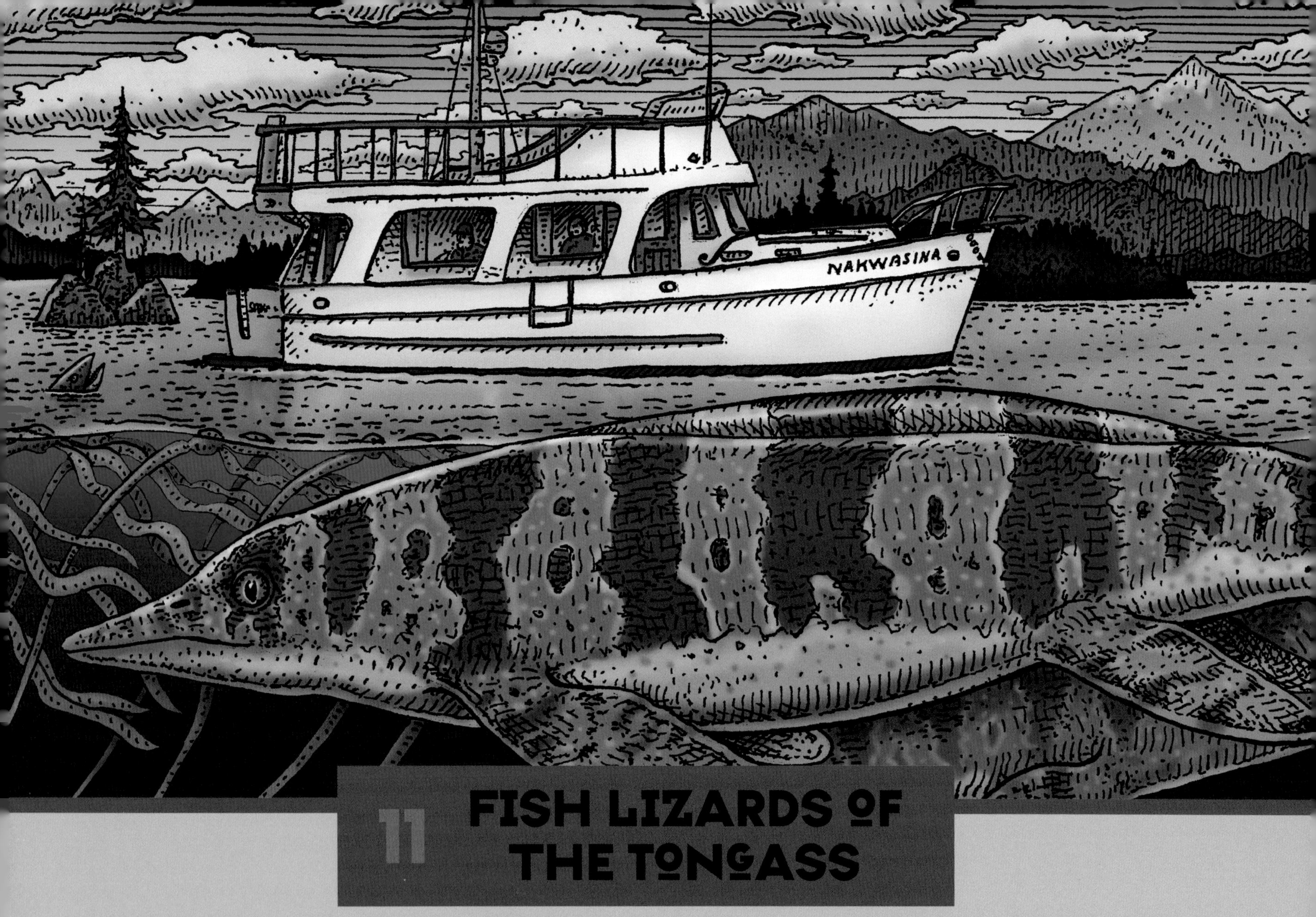


# 11 FISH LIZARDS OF THE TONGASS

For most of my life, I have collected and studied fossil leaves. As a result, I pay close attention to leaves and can recognize hundreds of different plant species just by looking at their leaves. Paleobotanists, the scientists who do this sort of thing, will often name new species of fossil plant based on their leaves and will publish journal articles illustrating the leaves. When you find a new fossil site and try to identify what you have found, you end up consulting these articles. This has been going on in earnest for more than a hundred years, and my office is full of books and articles about fossil leaves that date back to the 1880s and before. In 1994, I excavated a very unusual fossil leaf site in the town of Castle Rock, Colorado. The site looked like it was a fossil tropical rain forest, and it had more than 100 different types of leaves, many of which I had never seen before. I

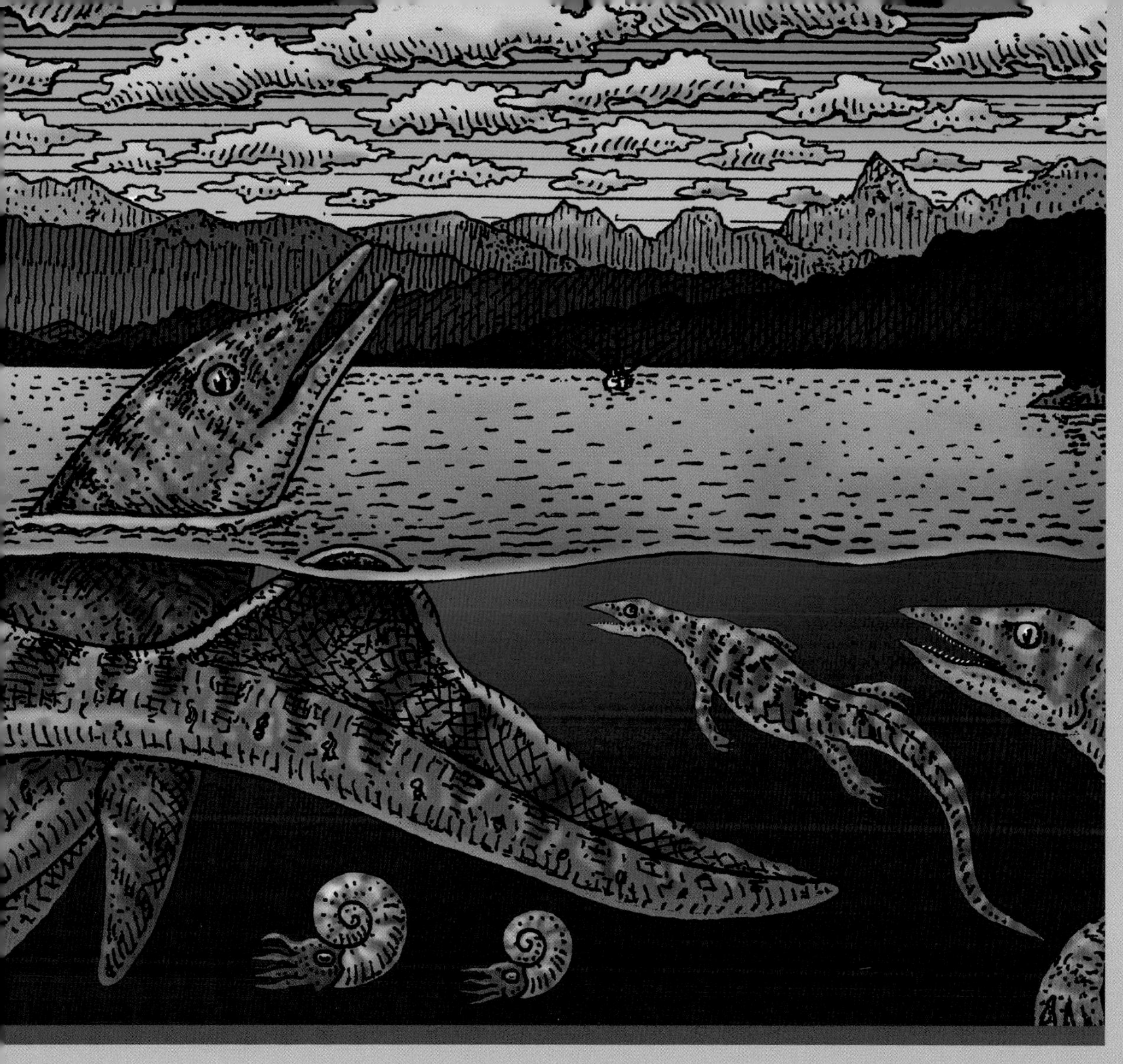

Looking out the window of the *Nakwasina* dreaming of a Triassic sea full of thalattosaurs, giant ichthyosaurs, and ammonites.

dug into the books to see if anyone had ever seen any of them before. After some months of searching, I came across a report written in 1936 by US Geological Survey paleobotanist Arthur Hollick that had several matches. His fossils came from Kupreanof Island in Alaska and were collected in 1907. I thought it would be interesting to return to that spot and see if the site also looked like a tropical rain forest. It didn't seem too likely since the site was in Alaska, but you never know.

I mentioned this to Ray and asked if he knew anyone who had a boat and would be willing to take us on an Alaskan leaf hunt. A couple of weeks later he called me back and said that he had a friend in Sitka who was always up for an adventure, and that if we helped cover expenses, he would take us where we wanted to go.

Ray and I landed in Sitka on a brisk and beautiful May day and were met at the airport by Barth Hamberg, a fit fifty-year-old landscape architect who designs trails and structures for the Forest Service. His boat was an elegant 37-foot cruiser called the *Nakwasina.*

Sitka is situated in an amazingly beautiful spot. Located on Baranof Island, it faces out to the open Pacific through an archipelago of small islands and is backed by high peaks to the east. Mount Edgecumbe, a massive snowcapped volcano, sits just 16 miles to the northwest and dominates the scenery. Sitka is the heart of Russian Alaska, and it wears its history proudly. Ray had been working on a painting of the Battle of Sitka and we went to visit the Sitka National Historical Park to look at the site of the 1802 fight between the local Kiks.ádi Tlingit and the Russians. This battle was an incredible example of the clash of cultures that unfolded in Alaska in the nineteenth century. The Russians had moved into Alaska because of the fur trade. When they came to Sitka, they brought Alutiiq hunters who traveled in kayaks called baidarkas. The local Tlingits were initially willing to coexist with the visitors, but eventually hostilities broke out and the Tlingits attacked the Russian fortress in 1802. After initially being repelled, the Russians returned in 1804, with a larger force that included a fleet of more than 400 baidarkas, and they besieged a wooden fort that the Tlingits had constructed. Ray's painting depicts one of the Tlingit counterattacks where the warriors are wearing wooden armor and helmets that are carved in the shape of sea creatures and scary men. These helmets still exist today in various museums, and Ray's image captures them in their moment.

Eventually, the besieged Tlingit, realizing they did not have enough ammunition to carry on the fight, slipped away at night and escaped by walking over the mountains to a place called Nakwasina Cove.

But that story is not the reason that Barth's boat is named *Nakwasina*. His boat is named after a famous canoe. The story of the *Nakwasina* canoe was told in the July 1933 issue of *National Geographic*. For their honeymoon in 1931, Jack Calvin (a friend of Ed Ricketts) and his bride, Sasha Kashevaroff, paddled the Nakwasina from Tacoma, Washington, to Juneau, Alaska, a distance of nearly 1,100 miles. They took their puppy, Kayo, and paddled about 20 miles a day from June 25

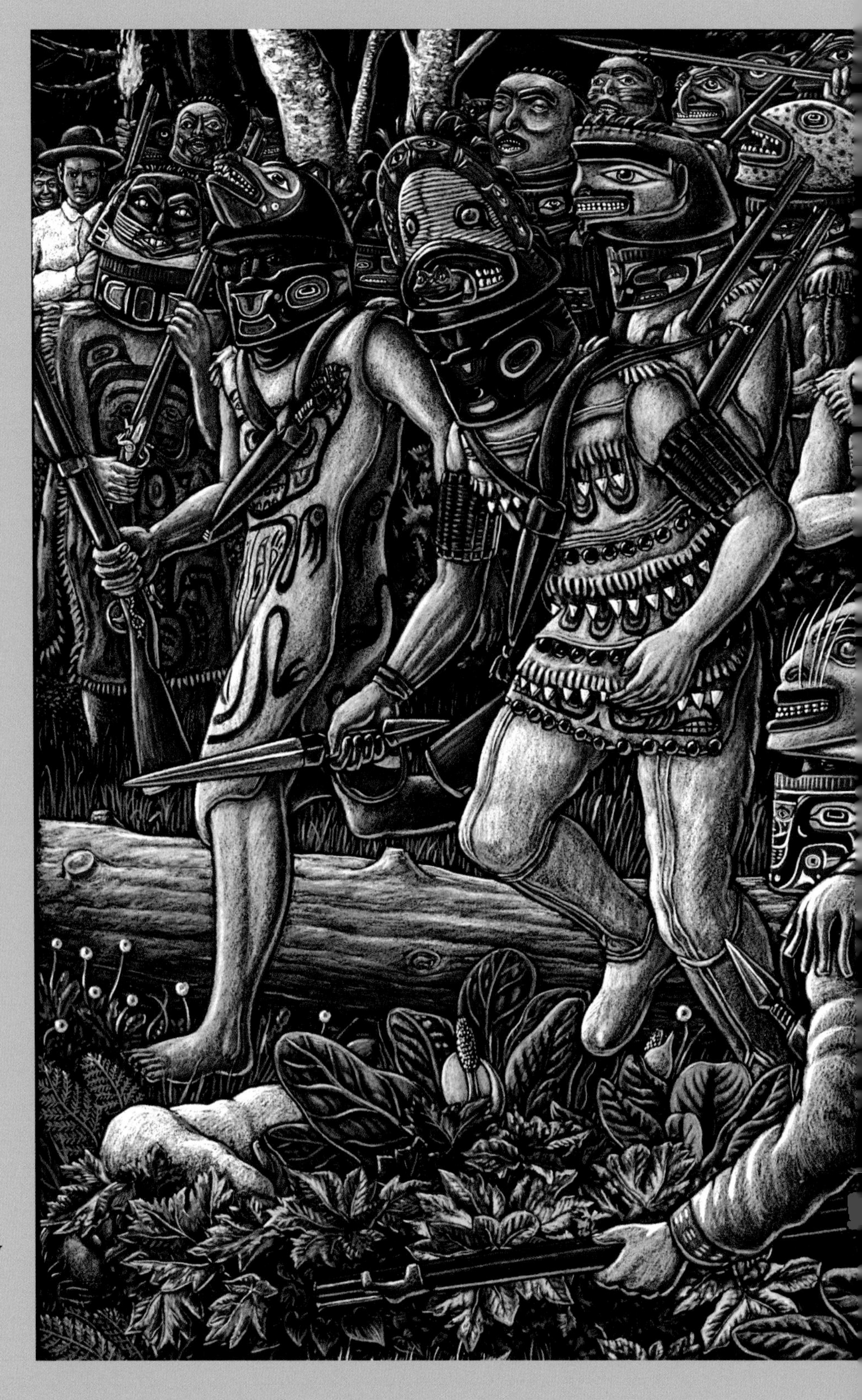

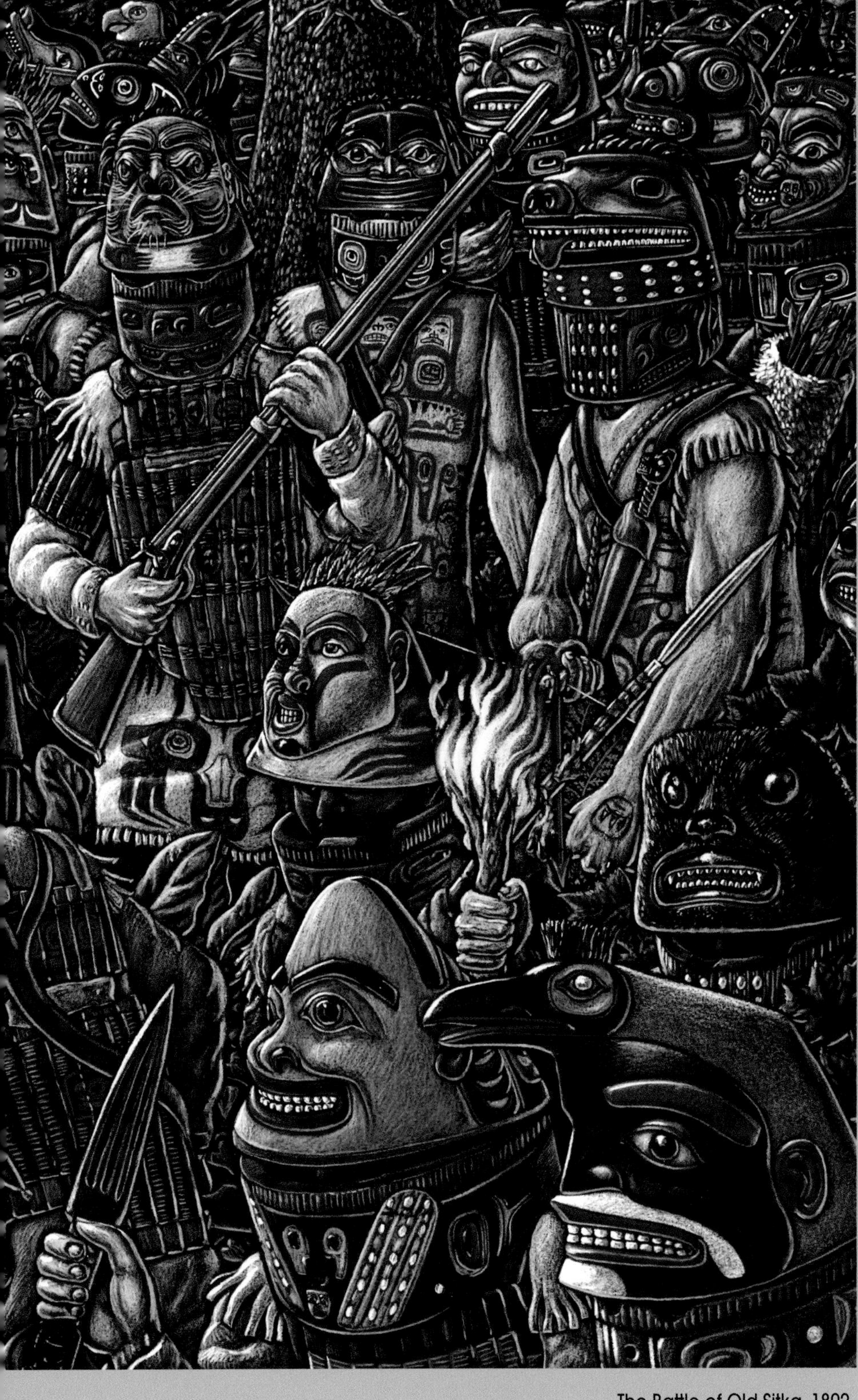
The Battle of Old Sitka, 1802.

to August 16, arriving fifty-three days after they started. Even by today's extreme sports standards, this was a heroic feat.

After their paddle, Jack and Sasha bought a small cruising boat they named the *Grampus.* In 1932, they met up with Ed Ricketts and Joe Campbell in Tacoma, and together they took the *Grampus* up to Alaska. This time the *Nakwasina* was strapped to the roof of the *Grampus*. Their voyage took them up Puget Sound and through the San Juan Islands, then through the entire inside passage and all the way to Sitka and finally Juneau. As usual, Ricketts collected marine invertebrates, deep conversations abounded, and, by all accounts, the group had a splendid time. Campbell was interested in mythology and religion, and Ricketts was growing more interested in the relationship of science to religion. And they all liked to party. The trip lasted two months, and for Campbell it was "epochal"; for Ricketts, it framed his ambitions to understand the ecology of the entire West Coast.

Ricketts and Jack Calvin published *Between Pacific Tides: An Account of the Habits and Habitats of Some Five Hundred of the Common, Conspicuous Seashore Invertebrates of the Pacific Coast Between Sitka, Alaska, and Northern Mexico,* which was finally published in 1939. Campbell published *The Hero with a Thousand Faces* in 1949. In 2015, Sitkans Jan and John Straley and Ed Ricketts's daughter, Nancy, connected these dots in their book, *Ed Ricketts: From Cannery Row to Sitka, Alaska.*

Jack and Sasha settled in Sitka in the mid-1930s, and, years later, Barth's *Nakwasina* was owned by Jack and Sasha's daughter, who named it after her parents' canoe. All of this meant that Ray and I would accidentally be traveling in a boat that was directly connected to Ed Ricketts and seeing the same waters that he saw 80 years later.

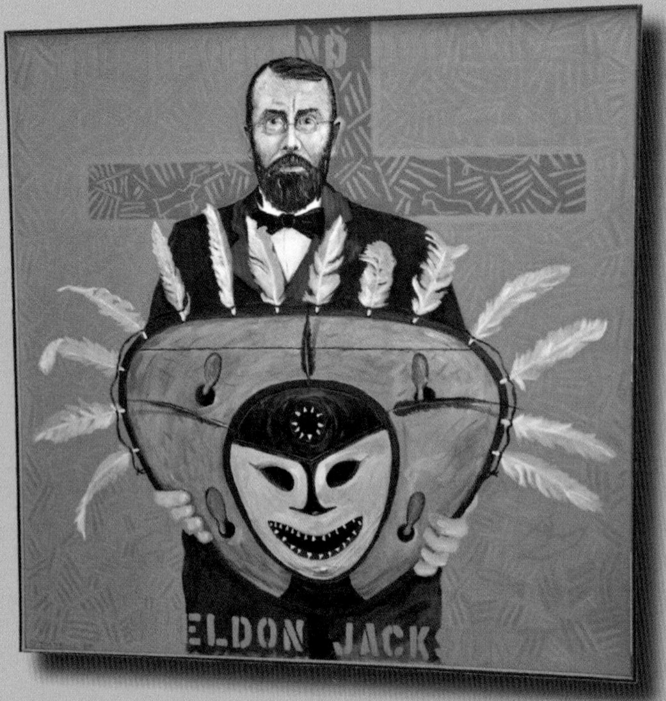

We spent a few days in Sitka, stocking up for our trip and seeing the local sites. I was eager to visit the Sheldon Jackson Museum. Jackson was an aggressively proselytizing Presbyterian minister who arrived in Alaska in 1877 and traveled widely in his search for souls. He founded many churches in remote Native villages and lobbied for the suppression of Native languages in favor of a Bible-inspired English. During his travels, he accumulated a remarkable collection of Native art and utilitarian objects, and, in 1888, he opened his museum in Sitka. The building and the collection are intact today and serve as a time capsule for the Alaska of 130 years ago. We stopped in to see Nadia Jackinsky and Jackie Hamberg, the curators of the collection, and they surprised us with a portrait of the Reverend Jackson that had been painted by Ray's brother, Tim Troll.

Tim had served as the city manager for the Yupik village of St. Mary's in the 1980s and had been very active in connecting local communities with objects made by their relatives and ancestors that are now in museums around the world. He organized the first exhibit of Yupik carved masks in Yupik communities. His portrait of the reverend is interesting because of the tension it expresses between who the reverend was and what he preserved by collecting. This small concrete building in Alaska is a microcosm of the relationship between Native communities and museums worldwide.

The collection itself was one of the best I have ever seen. The diversity of objects made from stone, wood, ivory, shell, skins, intestines, feathers, and metal attested to the human response to life in Alaska. This crossroads between two oceans and two continents has been a hotbed for creativity for literally thousands of years.

We also stopped in at the Sitka Sound Science Center, which houses a small aquarium with a strong research arm and an active program monitoring the humpback whales that migrate into Sitka Sound each year. Out on the sound, marine archaeologists were searching for the remains of Russian ships that had sunk after the battle of Sitka. In town, Tlingit artisans and Norwegian fishermen continue to make this tiny town of 9,000 one of the more interesting places on the West Coast.

The next morning, we packed the *Nakwasina*, slipped out of the harbor, and headed up Olga Strait along the west coast of Baranof Island. The *Nakwasina* moved at a stately 6 knots (or about 7 mph), so we were not headed anywhere fast and had time to scan the sea for whales and the shorelines for bears. After a few hours, we entered Peril Strait, a narrow and dangerous passage between Baranof and Chichagof Islands. One particular spot named Sergius Narrows had Barth quite focused. In this little pinch point, the wrong tide could spell a very bad outcome. Barth told us stories of flipped tugboats and crashed rescue helicopters, but he had also calculated our timing so that we were able to pass through at slack water.

At 10 a.m. we passed a channel called Deadman's Reach and glimpsed a little embayment called Poison Cove. It was at this spot that 150 of Baranof's Aleuts were hunting for otter when they stopped for a meal of clams or mussels. A red tide had contaminated the shellfish with a poisonous dinoflagellate, and the entire hunting party had died on the spot. Ray knew this story well because he had collaborated with a Ketchikan coffee roaster to make a brand of coffee called Deadman's Reach (Served in Bed, Raises the Dead).

By the afternoon, we had emerged from Peril Strait into Chatham Strait and could look across the 10-mile-wide channel and see Admiralty Island. We were directly across from the Tlingit village of Angoon, and I knew this name well. Paleobotanist Jack Wolfe had visited the Kootznahoo Inlet just south of Angoon in the 1960s and had found fossil plants that were 20 to 30 million years old. Barth had a different story about Angoon. In 1882, some of the local Tlingit were working for the Northwest Trading Company hunting whales.

Opposite: A portrait of the Reverend Sheldon Jackson painted by Ray's brother Tim.

Above: Barth Hamberg and his boat the *Nakwasina*.

An important Tlingit was killed in a whaling accident, and the tribe demanded reparations. The company appealed to the US Navy in Sitka, who responded by shelling one of the Native villages. I had never really thought about the naval engagements of the Indian wars. When I became the director of the Smithsonian's National Museum of Natural History, I was visited by an amazing group of Tlingit dancers from Angoon who were happy to hear that I had seen their village and were startled to hear about its fossil plants.

If you look at Chatham Strait on a map of Alaska, you will see that it is as straight as a ruler, a sure sign that the strait lies above a geologic fault. It is, in fact, one of the largest faults in Alaska. It trends north from Angoon before arching around to the north side of Mt. Denali and then continuing back to the southwest and out to the Bering Sea. Like the famous San Andreas Fault, the western side is moving north relative to the eastern side. The means that Baranof Island is moving north relative to Admiralty Island in the same way that Los Angeles is moving north relative to San Francisco.

We headed south down the channel and soon began to see humpback whales, Dall's porpoises, harbor seals, and otters. We spent the next day exploring the east side of Baranof and ended up anchoring late in the evening at the head of Red Bluff Bay. After dinner, Barth and I decided to go take walk and motored the inflatable dingy toward shore. As we approached the beach we spotted a brown bear in the distance. We entered the mouth of a small creek that was emptying into the bay and spotted two more bears. We landed the dinghy and climbed up onto the beach, taking care to keep an eye on the three bears. As we panned the shore with our binoculars, we spotted a mother bear with three cubs. With a total of seven bears in full view, we decided that it might not make sense to take that walk after all.

The next morning, we headed out across Admiralty Sound toward Kupreanof Island. This huge open body of water is famous for its feeding humpback whales who work as a group to trap fish in a net of bubbles. To do this, several whales dive straight down in a corkscrew pattern and

exhale as they dive. This creates a rising cylinder of air bubbles that surrounds and traps schools of fish. At the bottom of the dive, the whales turn upward and, as a group, swim up the center of

Above: A precarious rock formation on the shore of Admiralty Sound.

Right: Humpback whales feed by trapping their prey in nets made of bubbles.

the bubble cylinder with their mouths agape. It is a remarkable thing to see half a dozen 50-foot-long whales erupting from the calm surface of the sea. We were hoping to see this as we crossed the sound, but all we saw was the occasional whale spouting or fluking up before diving.

Toward afternoon, we approached a small group of islands off the north side of Kuiu Island. We decided to go ashore and check out reports of fossiliferous marine limestone. Barth nosed the *Nakwasina* into a narrow channel, and we found a safe place to drop the anchor. The tide was low, and about 9 feet of normally submerged shoreline was exposed. We rowed to shore and clambered onto the slippery rocks and starting looking for interesting marine life. The intertidal zone was extraordinarily rich and completely covered by a huge variety of kelp and seaweed. Technically, seaweed is marine algae, and it comes in a multitude of types and colors, including brown, green, and red. The rock walls that were exposed by the falling tides showed a remarkable zoning of different colors. In addition to the red, brown, and green algae, there were black, white, and orange lichens, and the whole shoreline looked like a stack of Neapolitan ice cream. Fortunately, Barth had a copy of Mandy Lindeberg and Sandra Lindstrom's *A Field Guide to the Seaweeds of Alaska* on board. We could see what had drawn Ricketts north and wished that he had lived to explore more of Alaska. Under the seaweed we found crabs, worms, snails, chitons, clams, and amphipods, and the place seemed completely untouched by pollution or people.

We hiked into the woods and walked on a forest floor that was like a thick sponge. Lichens, mosses, herbaceous plants, and ferns covered the ground so thickly that it wasn't possible to see the soil. Barth had taken us to see some really huge trees and eventually we made our way to a grove of immense Sitka spruce. These remarkable trees made me start to wonder about the geologic history of the Tongass Temperate Rainforest. When did it form? How was it affected by the ice ages? The answers to those questions were closer than I imagined.

We emerged out of the woods onto a rocky shoreline and realized that the only surface that was not completely covered with life was a narrow strip just above the high tide line. Here neither the marine nor the terrestrial life-forms flourished and it was actually possible to see the rocks. And the rocks were full of fossils. We crawled around and found crinoid stems and brachiopod

Right: Jim Baichtal's landing craft, the aptly named USS *Suspect Terrane*, made for easy access to the beach.

Opposite: Dawn Redwood (*Metasequoia occidentalis*) fossils on the shore of Kuiu Island.

shells. These rocks were Paleozoic, at least 300 million years old. Like the rocks at Kasaan and Glacier Bay, these fossils had formed not only a long time ago but also a long ways away. The emerging realization that drifting microcontinents carry with them their own fossil history is one of the many things that makes the interpretation of Alaskan geology so difficult and so exciting.

We reunited with the *Nakwasina* and motored over to the Tlingit town of Kake to refuel. Kake is home to a 128-foot totem pole, the tallest in all of Alaska. Like Angoon, Kake had been shelled by the US government, but this time it was the army in 1869, the same year that Seward had purchased Alaska from Russia. The Tlingit were not amused by the sale since it was their land to begin with and no one had bothered to ask them. Tensions brewed, fights broke out, people were killed, and cannons were fired.

After fueling up, we motored south along the coast. A few weeks earlier, we had planned a rendezvous with Jim Baichtal, and we radioed him while we were under way. He was camping on a nearby island with a small field crew. We pulled into a small cove and dropped our anchor. A few minutes later, Baichtal and two of his team came roaring into the cove in what can only be described as the perfect maritime fossil-collecting vehicle. It was an aluminum sled boat with a landing craft door for a bow and a deck full of tools. It was like an amphibious pickup truck. Baichtal had wryly named it the USS *Suspect Terrane*. Barth pulled out filets of white king salmon and began to grill them. Baichtal produced some steaks cut from a musk oxen that he had shot with a flintlock musket and began to regale us with tales of Alaskan geology.

It takes a while to get used to Jim because you initially think he's telling tall tales. But the longer you listen, the more you realize that his stories check out, and that he is an amazing guy who has direct access to the incredible geology of a large and remote place. It had been Ray's idea to include Jim on our expedition, and it was a very good idea indeed.

Jim had recently discovered a submerged volcano 10 miles offshore of Cape Addington on Noyes Island, about 100 miles south of where we had anchored. They found and mapped the volcano using sonar but had camera footage from submersibles that showed pahoehoe lava at a depth of 350 feet below the surface of the sea. This kind of lava only flows at the surface, so the volcano must have been at the surface when it was erupting about 15,000 years ago. This sounds crazy, but it makes sense in a world

where the sea level was rising and the land was sinking. Jim had used geology to prove that the coastline was much wider when the first humans made their way into North America from Asia. He also showed how rapidly the world can change.

It was cloudy when we woke up, and for some reason we had chili for breakfast. Jim and his crew returned from their camp, and we all hopped into Jim's sled to go off in search of a fossil site that had last been visited in 1907. Both Jim and I had read the original reports, and they were pretty vague as to where we might find the fossil leaves. Even if we found them, there was no guarantee that I would find the ones I was looking for.

The good thing about geology is that there is a certain level of predictability. We had increased our odds by timing our arrival with a very low tide so we could see all of the rocks that the guys back in 1907 had seen. As we motored along the coast, we noticed that the falling tide had exposed a flat terrace of bedrock. That was a good sign, but the other thing we noticed was that there were large wheelbarrow-sized granite boulders lying on the terrace. These were obviously glacial erratics (boulders that were transported into an area by a glacier or ice sheet and were left lying on the landscape after the ice melted), but it meant that boating was going to be quite hazardous unless we moved very slowly.

Jim nosed the boat onto the beach, and we ran off and started cracking rocks to see if there were fossils. The bedrock exposed on the beach was sandstone, shale, and conglomerate, which was a good sign because it meant fossil leaves were possible. A few of us found some carbonized fossil logs, which was even better news since it meant there were plants here. After an hour of not finding any fossil leaves, we got back on the boat and moved down the beach and tried again. By now it was raining, but we were in Alaska so it didn't matter. We tried another spot with no luck and kept going.

Then Jim spotted a mink on a log up at the top a beach and thought that seemed like a good sign. We pulled ashore again and scattered across the wide beach. Up by where the mink had been, I found some shale that had some poor leaf impressions. A good sign, but not what we were looking for.

1. *Shonisaurus* sp., large ichthyosaur
2. *Toretocnemus californicus*, small ichthyosaur
3. Unnamed small thalattosaur
4. *Hybodus* sp., shark
5. Ammonite
6. *Saurichthys* sp., bony fish

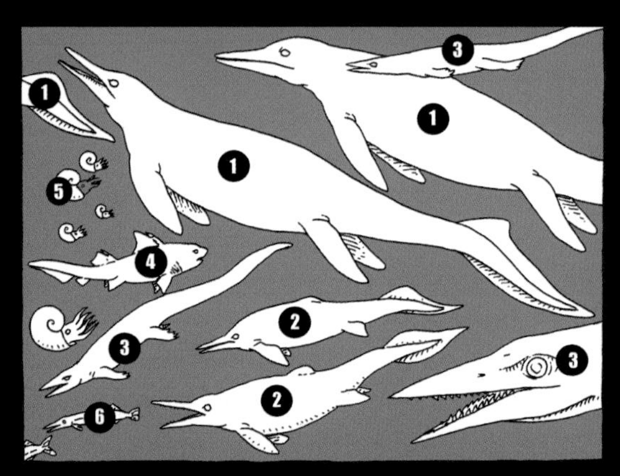

Jim and Barth were working down near the water's edge, and Jim was using a big hammer to move and break a lot of rock. It started raining harder. Then Jim yelled, "Is this what you are looking for?" Music to my ears, and I rushed down to the shore where he had split a flat rock the size of a dinner tray. The exposed surface of the rock was covered in *Metasequoia* leaves – the very same kind of conifer that I had found in Republic, Washington, in 1977.

Now it was pouring, but I was really excited and we started pulling up slab after slab and the leaves started coming. I split a big slab and there, right in the middle of it, was a 10-inch-long complete leaf of the same species that I had found in the Castle Rock, Colorado, site and that had also appeared in Hollick's 1936 report. We had found the right spot.

We kept splitting rocks and finding more fossils. I had been hoping to find fossil cycads, and we found them as well. I could not believe my luck. By now, the tide was flooding and time was running out, so we packed up our fossils in milk crates and loaded them onto the USS *Suspect Terrane*. We departed with me knowing full well that I would return to this place.

We spent the next few days prospecting the areas for new sites. While we found several, none were as nice as the site by the mink on the log. We saw black bears, porcupines, Sitka black-tailed deer, and an incredible number of sea otters. We counted more than 200 rafts of otter before we stopped bothering. Jim called them sea weasels and noted that their arrival in this bay had meant the end of the Dungeness crabs. Otters are a keystone species, and their presence usually has a dramatic positive effect on the ecology of the kelp forests because they eat the sea urchins that eat kelp. But since people like to eat Dungeness crabs, the sudden growth of a local sea otter population is sometimes not welcome.

Landing on Hound Island.

At the end of the second day, we motored over to Hound Island, a small tree-covered island where Jim and his crew had been camping. They were prospecting for Triassic marine fossils in a dark-gray shale that was exposed in the intertidal zone. Ray and Jim had been here before, and Jim had found enough individual bones to lure scientists from Dallas, Seattle, and Fairbanks to this small, remote island. They had recovered parts of a *Shonisaurus*, a whale-sized marine reptile, and on his last trip, one of his team had found the back end of a small 10-inch-long marine reptile. He sawed out the slab of rock that he hoped would contain the skull and shipped it to Fairbanks. The guys in Fairbanks hired fossil preparator J. P. Cavigelli from Casper, Wyoming, to fly to Alaska to use needles to see if he could find the skull. I looked at the slab in Fairbanks and deemed the task hopeless. But I was wrong. J. P. teased out a perfect skull with a needle-nosed snout. With the complete skeleton exposed, it was clear that the skeleton was a new and unknown species of a group of animals known as thalattosaurs.

After four days of digging in the rain, we headed back toward Sitka and Jim took off to his home in Thorne Bay. Before he left, he gave me the coordinates of a fossil site we could check out on the way back. Barth steered the *Nakwasina* back out into Frederick Sound and around the north end of Kuiu Island. The sun had finally come out and it was a splendid afternoon. I was feeling really good about finding the fossils near the mink log.

As we cruised slowly along the forested shoreline, I scanned the cliff face for possible fossils. The three of us were up on the top deck enjoying the sun and enjoying life. We could see for miles, and there was not another boat around.

As I glanced at the cliff, I caught a glimpse of something that didn't look quite right. I trained my binoculars on it and to my amazement, I was looking at a red ochre face. It was a Tlingit pictograph of a face surrounded by jagged rays high on the limestone wall. Neither Ray, Barth, nor I had ever seen anything like it. It did not look too welcoming.

Barth idled the engine, and we sat floating about 40 yards from shore and thinking about our luck. Just then, a humpback whale surfaced between us and the shore. It had come up from behind us and was intent on feeding right up against the cliff. As we watched, it would dive again and again, always surfacing within 20 feet of the rock wall. Barth slipped the engine into gear, and, being mindful to keep our distance, we kept pace with

the lazily feeding whale. For the next forty minutes we shared the afternoon with the whale. Finally we pulled away and headed toward the small embayment that Jim had pointed me toward. As we entered the bay, I scanned the shore for the feature Jim had described. Just then, a small black bear sprinted across the beach and up into the woods.

Barth dropped the hook and we rowed ashore. The entire beach was a 350-million-year-old coral reef composed of a variety of corals, including the solitary horn corals and larger masses of colonial coral. We walked around in the lengthening evening light, amazed by this ancient reef and the odd circumstances that had placed it on a modern shoreline where it was encrusted with barnacles, mussels, and seaweed. It seemed so odd that a 350-million-year-old shoreline would once again be a shoreline. That's when it really hit me about the antiquity of the West Coast, the eternal coastline.

It was a really nice way to end the day and the trip. The next day we motored all the way back to Sitka. At the Sitka airport, everyone carried the classic Alaskan fish boxes full of salmon and halibut. I had fish boxes too, but mine were full of fossil leaves.

Above: The Tongass palm forest of the Paleocene.

Opposite: Mississippian-aged fossil coral covered with modern barnacles.

Above: Jim Baichtal using a diamond saw to loosen up the bedrock of the beach.

Right: Jim Baichtal, Kirk, and Ray celebrate the discovery of the Kupreanof palm frond.

Twenty months later, Ray, Jim, Barth and I returned to Kupreanof, and this time I brought a film crew from London. We were filming "Making North America," a three-hour *NOVA* special that was a biography of the North American continent. The crew and I had just been filming on top of the Juneau Icefield and had flown to Kake. Barth, Ray, and Jim arrived by boat. We came to film fossils in order to make the point that the West Coast of North America was, in part, composed of suspect terranes from elsewhere in the world.

We returned to the mink log site at low tide and started cracking rocks and finding fossils. It was a sunny day, and the film crew was getting decent footage, but the fossil leaves weren't showing up that well on the camera and the director was getting a bit frustrated. Then I noticed something. It was a little piece of palm leaf about 4 inches long. If there was a little piece of palm, then it was possible we could find a whole palm leaf. Knowing that you can't find a big leaf on a little rock, I urged the guys to pry up larger slabs. They worked hard and we split a lot of big slabs, but we were coming up with "big nothing." With the tide coming up, we worked harder. Ray and Jim were prying away at an outcrop that was only a few feet from the advancing tide. I looked at it closely and realized that they had the edge of a palm frond. We all got together and drove chisels into the rock. Finally, we slipped a crowbar into the crack and, on the count of three, we heaved and pried the heavy slab up.

Much to all of our utter amazement, the slab contained the central portion of a huge palm frond. We found it, on camera, and minutes before the tide flooded the outcrop. Again, I could not believe my luck. We quickly carried the slab up the beach and then pulled out two adjacent pieces of equal size. When all was said and done, we had found a palm frond that was 7 feet wide!

Ray and I both thought about the palms of Chuckanut Drive near Bellingham and this palm on Kupreanof Island and started to wonder about the climate implications. I cannot imagine a better way to describe global climate change than by uttering these two words: "Alaskan palm."

We boated away from the site and headed south toward Prince of Wales Island. Well out into open water between two islands and about 2 miles from shore, we passed a couple of seals. But something about those seals seemed wrong to me. I didn't recall that seals had big ears. I yelled to Jim to turn the boat around and have another look. When we got close, we realized that they weren't seals at all, but a pair of swimming wolves. Embarrassed, the wolves rapidly swam to the nearest island and disappeared into the forest.

We continued south so that Jim could show us the karst caves full of bear skeletons, and all I could think about were submerged volcanoes, Alaskan palms, and marine wolves.

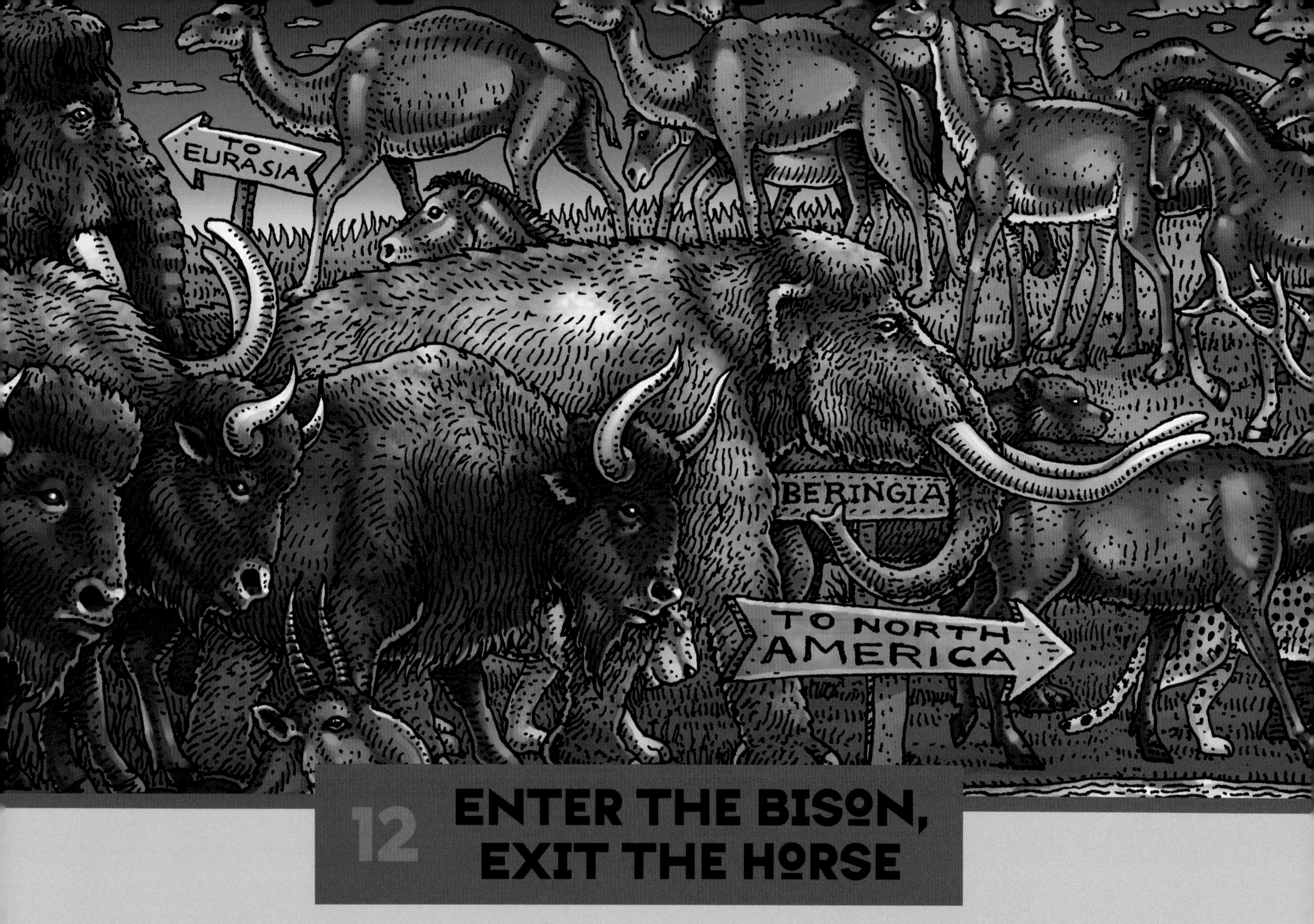


# 12 ENTER THE BISON, EXIT THE HORSE

Alaska should really be at least a dozen states. It's that big, and the various parts are very different from each other. There is the forested maritime southeast, the glaciered southern coast, the remote and wind-blasted Aleutian Islands, the treeless North Slope, the Brooks Range, the Bering Sea coastal plain, the volcano-rich Alaskan Peninsula, the Yukon Valley, Kodiak Island, the Anchorage–Wasilla urban corridor, the Denali country, and lots of other unique places I haven't named.

There is a really important part of Alaska that is presently underwater. It's called Beringia, or the Bering Land Bridge, and it was once a continuous land connection to Asia. Today, it is covered by the Bering and Chukchi Seas. The Bering Sea is a shallow water extension of the Pacific Ocean, and the Chukchi Sea has the same relationship to the Arctic Ocean. Both seas are shallow, less than 200 feet deep in most places. In the winter, the seas freeze over and are covered by floating

For millions of years the land bridge connecting Siberia and Alaska acted as a biological pathway between Eurasia and North America.

sea ice. The ice melts and breaks up in June, and the water is open until late September. Twenty thousand years ago, when glacial ice filled Puget Sound and much of Canada was covered by a vast continental ice sheet, global sea levels were 300 feet lower than they are today. If you subtract 300 feet of sea level from a sea that is 200 feet deep, you have dry land, and that is what the northern Bering Sea and the southern Chukchi Sea used to be. When these two seas receded, a 1,000-mile-wide land connection between Asia and North America emerged where there had once been saltwater. When the continental ice began to melt around 12,000 years ago, sea levels rose, flooded the land connection, and eventually formed the Bering Strait, which connected the Pacific Ocean to the Arctic Ocean.

Not only did many species of plants and animals make use of this connection to migrate from Asia to North America or from North America to Asia, but this is almost certainly the path that brought the first of many migrations of people to the Americas.

When I was working at the US Geological Survey in California, I met David Hopkins, who brought real science to the understanding of Beringia. He was a scientist with extremely broad interests, and he was particularly good at synthesizing information from many fields. He began to study the Bering Land Bridge and brought together the insights from geology, paleontology, archaeology, botany, and other fields. He also realized that Alaska was only half the story of Beringia and so forged working relationships with Russian scientists who were working on the other side. In reference to David's huge influence on the understanding of this region, Dan O'Neill wrote a book about him titled *The Last Giant of Beringia*.

While David largely focused on the Bering Land Bridge of the ice ages, other scientists were beginning to realize that the connection between Asia and North America was very old and had influenced the composition of North American plants and animals not just for tens of thousands of years but for tens of millions of years.

The similarity of Cretaceous dinosaurs in central Asia and western North America emerged slowly over the twentieth century. Roy Chapman Andrew's famous expeditions to the Gobi Desert in Mongolia in the 1920s had discovered *Protoceratops* dinosaurs that were clearly related to *Triceratops* and other ceratopsians from Canada and the United States. Even earlier, Russian discoveries of flat-headed hadrosaur dinosaurs from the China–Russia border region of the Amur River had suggested a relationship between the two continents. Later expeditions to the Gobi Desert by the Russians in the 1940s and the Poles in the 1960s discovered a host of other dinosaurs that had close North American relatives. These included the Mongolian *Tarbosaurus*, which was a close cousin to the American *Tyrannosaurus* and the crested hadrosaur *Saurolophus*, which occurred both in Mongolia and in Montana and Alberta. I went to Mongolia and China between 1997 and 2008 and found several species of Cretaceous fossil plants that I recognized from North Dakota and Montana.

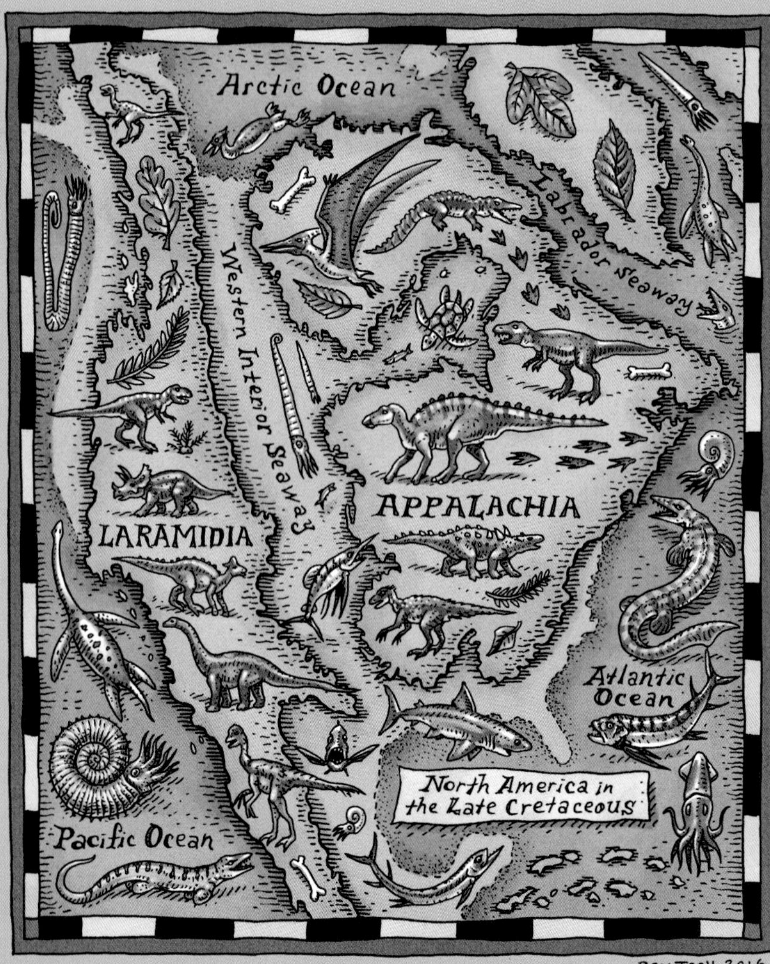


By the 1990s, it was well established that the connection between Asia and North America had been intermittently open as far back as 100 million years. In fact, for much of the Late Cretaceous when the Western Interior Seaway split North America into two subcontinents (Laramidia to the west and Appalachia to the east), western North America looked like a long, drooping peninsula of Asia.

But because western Alaska today is hard to access, few people realize its critical role as a gateway between two continents. And Beringia is not just a story of land migration. When the land bridge was intact, it also formed a barrier between the North Pacific and the Arctic Ocean. Today the Bering Strait blocks land animals but allows marine animals such as gray whales and narwhals to move from one ocean to the other.

Ray and I realized that to tell this story, we needed to go to Fairbanks, home of the University of Alaska and the Museum of the North. We also realized that to tell the story accurately (and to earn a few bragging points), it would be important for us to go there in the dead of winter.

It was a calm January day when we landed at the Fairbanks airport and rented a car. The temperature was a typical –32°F, and a thick snow draped the trees and ground. We stopped at a coffee shop to warm up and found that the baristas were wearing shorts and Hawaiian shirts. People of Fairbanks are proud of their weather and unbowed by it.

We were in Fairbanks because we wanted to learn about Beringia during two completely different time periods. The Beringia of the mammoths was one that occurred when the world's climate alternated between ice sheets in Seattle and ice sheets in Alaska. This is the icehouse world. The Beringia of dinosaurs was a world without ice sheets, a hothouse world. By coming to Fairbanks we could meet people who studied both of these worlds.

The Museum of the North sits high on a hill on the University of Alaska campus and is the only natural history museum in Alaska with a full research, exhibit, and education mission. It first opened in 1929, and was rapidly stocked with Native Alaskan artifacts, biology specimens, and fossils collected by the exuberant Otto Geist.

The permafrost around Fairbanks contains placer gold that was deposited by streams and rivers before the landscape began to freeze. The first gold miners to the region rapidly discovered that they had to cope with the permafrost if they wanted the gold. The answer was relatively simple: spray pressurized water onto the frozen sediment to melt it, and then run the resulting slurry through a sluice and extract the gold. They got gold but they also got something that they did not expect: a bycatch of well-preserved bones and tusks from Ice Age animals.

Otto Geist knew an opportunity when he saw one, and he began harvesting bones from the gold mines. Stories of his discoveries reached the ears of Childs Frick, a New York philanthropist and vertebrate paleontologist who worked with the American Museum of Natural History. Frick started funding Geist, and the museum began to receive a steady stream of crates full of Ice Age bones.

When the Smithsonian's National Museum of Natural History began to build its hall of Ice Age animals in the early 1970s, curator Clayton Ray went to New York and selected enough bones from Geist's collection to fill the entire hall. Meanwhile, back in Fairbanks, Geist continued to collect, filling room after room on the campus with fossils. There is even a rumor that Geist ran out of space and buried a cache of mammoth bones and tusks somewhere on campus. Geist passed away in 1962, and the exact spot of the cache remains a mystery.

As Ray and I drove up the hill toward the museum, we spooked a large flock of sandhill cranes that flew off into the icy sky. We crested the hill and saw the new museum, which opened in 2005. Designed by

Way up North, blasting gold and fossils out of the "Loess," with giant water cannons...

architect Joan Soranno, the building is a series of swooping white panels and walls meant to represent the landscape of the far north. The design is a stunning success, and the building merges seamlessly with the frigid landscape it occupies. Perched on the edge of the hill, it has a broad vista of the Tanana River Valley and the distant Alaska Range.

The museum itself is full of natural history and art of the far north. We slowly wandered through the halls looking at paintings, sculptures, Native artifacts, mammoth tusks, walrus skulls, fossils, seal skeletons, and caribou antlers. One specimen that I had longed to see for years was the 36,000-year-old mummy of an Ice Age bison (*Bison priscus*) known as Blue Babe. We rounded a corner and there it was, a complete bison mounted with its legs tucked underneath it as though it was resting. Its skin was indeed blue, a function of a blue mineral called vivianite that grows on certain organic remains in the permafrost. While we were reveling in the proximity of the magnificent beast, we overheard two women and realized that one of them was the granddaughter of the miner who found Blue Babe. This is truly one of the great fossils of North America. The place was magical, and the best part was still to come: we were there to see the research collections and to interview Pat Druckenmiller and Dale Guthrie. Pat

Blue Babe, the mummified steppe bison (*Bison priscus*) on display at the Museum of the North in Fairbanks.

was hired in 2007 to replace the retiring Roland Gangloff. Dale had been in Fairbanks for most of his career and had recently retired from his Ice Age studies to take up painting and sculpture.

Pat greeted us and took us down to his tidy office in the basement. He is a fit fellow with a well-trimmed beard that exuded a sense of hardy efficiency. He grew up in Colorado, the son of a wildlife biologist, and he studied paleontology in Wisconsin and Montana. His specialty was marine reptiles, but he was good with dinosaurs as well. As we talked, I learned that he had been part of the Montana team that prepared the *Tyrannosaurus rex* skeleton that I had just borrowed to become the centerpiece of the new Smithsonian dinosaur hall. I had been reading about Pat's recent fieldwork in Svalbard, Norway's Arctic Islands, where he was part of a team collecting Jurassic ichthyosaurs in an area teeming with polar bears. Now that he was based in Alaska, he had started to explore the distant reaches of the state in search of Mesozoic reptiles. He and Ray had crossed paths with the Gravina Island ichthyosaur dig, and they had recently worked together on the exquisite little needle-nosed thalattosaur skeleton from Hound Island. He told us about a headless Triassic ichthyosaur that had been found in the Brooks Range in the 1960s, and how the Fairbanks team had figured out how to excavate it in one huge slab and extract it by helicopter in 2002.

Pat's predecessor, Roland Gangloff, had worked with Bill Clemens and the UC–Berkeley team to do the first excavations of the polar di-

Upper: Pat Druckenmiller and the cast of an ichthyosaur.

Right: Kevin May with the limb bone of a duck-billed dinosaur.

nosaurs on the North Slope. In 1961, A geologist named Robert Liscomb found a layer of bones along the Colville River and collected a few thinking that they were Ice Age fossils, never even considering they might be dinosaur bones. He was killed in a rock fall the next year, and the bones lay hidden in a sample room at Shell Oil until the early 1980s, when USGS paleontologist Charles Repenning (of *Neoparadoxia* fame) saw them and realized they were dinosaurs. This was the realization that sent Bill Clemens and the UC–Berkeley crew to the site in 1982. They made several more trips during the 1980s, and then Roland Gangloff continued the excavations into the 1990s. Tony Fiorillo started working here with Roland Gangloff in 2000, and Pat showed up in 2007. Collectively, they have found an amazing array of species that includes representatives of all of the major groups of Late Cretaceous dinosaurs.

Ray and I were hoping to join Pat on his next trip north, and to our delight, he readily embraced the idea and invited us to join him the following August. The conversation turned to what we would need to bring, how we would get there, and bears. For all of his time in Alaska, Ray has only just begun to quell his quite reasonable fear of bears. Pat described the guns and electric wire fences that we would use while camping on sandbars. He didn't seem to be worried in the slightest.

He then took us into the collection vaults where we met Kevin May, the head of operations for the museum. Kevin is a wisecracking cinder block of a guy who looks like a football center. Like me, he grew up in Washington State hunting for fossils, and now he managed the museum at the top of the world.

THE MAMMOTH TWO STEPPE

The collections were remarkable. We wandered the packed hallways, pulling open drawers of dinosaur bones, ammonites, clams, fossil leaves, teeth, and skeletons from a time when Alaska was much warmer. The tops of the cabinets were covered in mammoth tusks, bison skulls, horse and camel bones, and other treasures from the Ice Age. They had great specimens of some of the rarer animals such as the American lion, the giant muskoxen (*Bootherium*), and the Ice Age moose (*Alces*). This museum is an icehouse-hothouse treasure trove.

Ray and I had lined up lunch with Dale Guthrie, and the three of us met in a tiny Thai restaurant in old-town Fairbanks. I had wanted to meet Guthrie since I first read his 1990 book, *Frozen Fauna of the Mammoth Steppe: The Story of Blue Babe*, which tells the story of the bison mummy's discovery in 1979 and all the work that Dale did to reconstruct the animal's life and death.

In a really neat piece of work he showed quite clearly how the bison had been killed and partially eaten by an American lion. The book also expanded on his hypothesis that the Bering Land Bridge had once been covered by a grassy steppe rather than the tundra that grows along the shores of the Bering coast today. He called this the Mammoth Steppe, and in so doing kicked off an ongoing debate about the very nature of the vegetation of Beringia and its potential to support large populations of animals. His idea was that as the sea rose and flooded Beringia at the end of the Ice Age, the local weather grew wetter and wetter, eventually destroying the Arctic steppe and dramatically decreasing its ability to support a large population of mammals. Sometime between 14,000 and 10,000 years ago, the mammoth, bison, and horse became extinct in Beringia, and the humans and moose moved into North America from Asia.

I always find it fascinating to meet scientists whose work I have appreciated, and I was eager to get to know the man. Dale did not disappoint. He is a tall and intense person, with strong and provocative opinions and a way with words. He is balding and gray with a tightly cropped beard, but he exuded fitness and energy. He was happy to meet Ray, whose art he had seen and admired, and he told us how he had retired from paleontology to become an artist himself. Over a lunch of Alaskan pad Thai, Ray, Pat, and I chatted with Dale about the story of Blue Babe and the Mammoth Steppe.

Dale was born in Illinois, raised as a Mormon, and initially thought he would become a priest, but he ended up with a PhD from the University of Chicago in 1963 instead. His first faculty appointment took him to Fairbanks and launched him into Beringia. He was just about to leave for a sabbatical in France in 1979 when gold miner Walter Roman's hydraulic jet exposed a mummified bison carcass in his mine. Academics treasure their sabbaticals, and Dale was annoyed to have his derailed by a fossil. But he knew that he was looking at a once-in-a-lifetime find, and he got down to business.

The first thing to emerge from the frozen muck was a hoofed foot. Dale said it looked like a Fudgesicle. Bit by bit, more emerged. Then the tail fell out by itself. The hide was attached to dried meat that looked like beef jerky. As the pieces

Dale Guthrie and his painting of the fifteenth-century B.C. Battle of Megiddo.

emerged, they were stored in the museum's freezer, and pretty soon, Dale had the whole carcass. This was not the first Ice Age mummy from the region. In 1948, miners found a baby mammoth mummy they named "Effie," but this little guy ended up on display at the American Museum of Natural History in New York. It was clear that the bison mummy was going to be a treasure for the museum in Fairbanks.

Dale then called in Björn Kurtén, the great Ice Age mammologist from Finland, to help him figure out how to mount the specimen; they consulted with Russians who had mounted frozen Ice Age mummies from Siberia. Dale told us that one of the people they sent over had been trained by the guy that "pickled Lenin." The team worked long and hard to reassemble the frozen bovine Humpty-Dumpty and eventually ended up with the splendid carcass on display at the museum today.

Some paleontologists like to taste their fossils, and there was a rumor that a bit of Blue Babe ended up in a stew. When we asked Dale what Blue Babe tasted like, he responded "mud." In the end, Fairbanks had an iconic fossil and Dale had a book and a new theory. Once the Blue Babe smoke had cleared, Dale got back to his sabbatical, which involved the in-depth study of Paleolithic cave paintings. He had thought long and hard about what it was like to be one of the first humans and how they had looked at their world.

Dale invited Ray and me to visit him at his house, and later that afternoon we drove out beyond the western edge of town and past a musk-oxen farm to his house in the woods. The man who greeted us was an artist and a hunter, not the scientist we had met at lunch. His home was full of bronzes, wood carvings, paintings, and the skulls of many trophy mountain sheep. Some of his subject matter was relevant to his life as a paleobiologist, and we admired bronzes of mammoth, mastodon, and bison. His more recent work was of a much more classical style and subject, as he felt that the only way for him to learn how to paint was for him to copy classic paintings. He had just finished a rendering of Gustave Caillebotte's *Les raboteurs de parquet* that was stunningly similar to the original. His active canvas was a huge panoramic battle scene showing the

## Desmos on My Mind and Under a School

You may have noticed a lot of desmostylian-inspired artwork in this book. I've been smitten by these wacky beasties for a long time now and have filled up a few sketchbooks with renderings of them. I remember Kirk casually mentioning them twenty years ago as something I ought to check out. My interest was piqued even more when I met Pieter Folkens in the mid-1990s. Pieter's a well-known painter of marine mammals and was spending his summers on cruise ships as a whale-watching guide. Pieter came by my studio a few times, and we soon started diving into the topic of prehistoric marine mammals of the West Coast. I was working on an illustration of an extinct giant Steller's sea cow at the time, and Pieter had some helpful insights on what this massive beast had looked like. We decided to build a small clay model of one to better understand the proportions. Every week when Pieter's ship would come to town, we'd get together for a cup of coffee and to tweak the model.

One week Pieter loaned me a small stack of scientific papers on desmostylians and encouraged me to check out this group of mysterious long-vanished marine mammals. I was intrigued and copied the papers and started a folder on them. The "desmo" brain worm began to take root in my gray matter.

So what's the big deal about desmos? Why shouldn't you too feel the warm glow of Desmophilia? I'm drawn to the offbeat and bizarre, and these critters fit the bill nicely. The word "Desmostylia" is Greek for "bundle of pillars." The name describes their wacky-looking teeth that resemble small six-packs of beer. They ground these teeth away completely during the course of their life.

They're big in Japan where they have an almost pop-icon status. You can buy plastic desmo toys, lunch boxes, and comics – or even attend theme parks and play on gigantic life-sized sculptures of them. But here in the United States they're virtually unknown by the general public. Kirk and I vowed to change that.

Because many of their bones are so oddly shaped and proportioned, Japanese, American, and British scientists have fiercely debated their walking and swimming postures for decades. One scientific paper would postulate a frog-like stance, only to be answered a few years later by a tippy-toe stance. One of the latest papers from Japan advocates a theory that has them crawling along on their enormous bellies. All of the scientists do seem to agree that they swam like polar bears with a sort of dog-paddling motion.

I recently spent some time with Howard University's Daryl Domning in the collection rooms of the Smithsonian Museum of Natural History. Daryl laid out a skeleton of a desmo on the museum floor and

Daryl Domning with a cast of a *Desmostylus* skull from Japan.

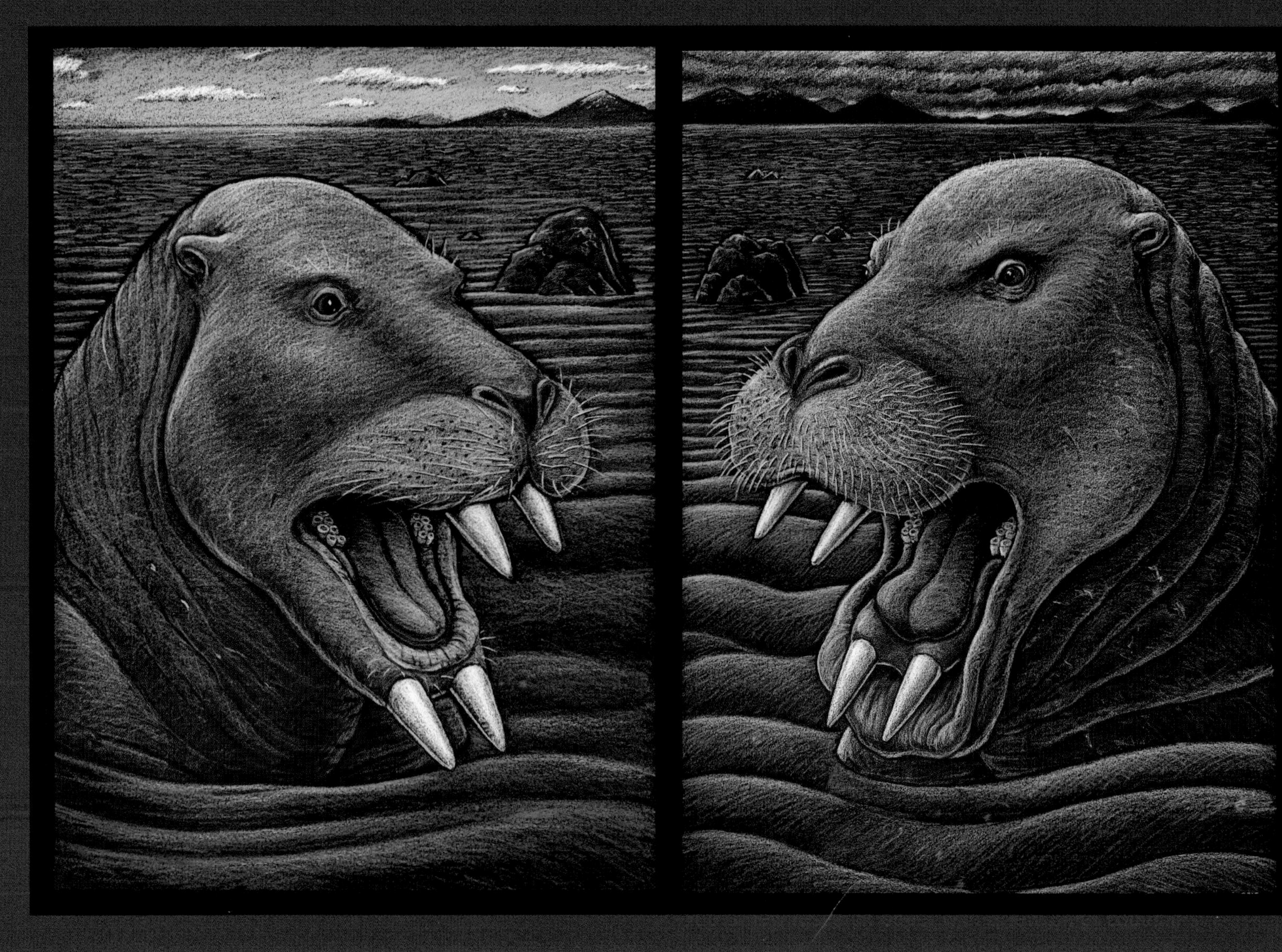

Two alternative reconstructions of *Ounalashkastylus*, one with lips (right), and the other without (left).

explained how the bones fit together. I became a believer in Daryl's upright stance theory at that point, but I'm betting the debate isn't over yet.

Their evolutionary origins are another bone of contention. They look a lot like hippos at first glance, but for many years most scientists thought they were an offshoot of the elephant family. The latest analysis suggests that they are kin to the rhinos and horses.

In 2015, I met Lou Jacobs and Michael Polcyn from Southern Methodist University at a geology conference in Anchorage. Lou and Michael were finishing up a new research paper with a few other coauthors on an exciting desmo discovery made years before in Unalaska out in the Aleutian Islands. A new big desmo from Alaska! I was thrilled, and offered to do some drawings for the press release announcing the discovery.

For nearly 60 years, experts have argued about desmostylian posture. Here are eight hypotheses.

Back in 1950, several large bones were found in a rock quarry where an elementary school was later built. The bones were kept in a Native corporation office for fifty-four years until Dallas Museum of Natural History paleontologist Tony Fiorillo ended up in town and gave a talk about his work in Alaska. Some local showed Tony the bones, and he immediately realized their importance. It turned out that there were bones from four individual desmostylians, ranging from juveniles to adults. They realized they'd found a 23-million-year-old sample of a breeding colony from the ancient coast of Alaska – a rare find indeed. It's the northernmost desmo find as well.

They named the new animal genus *Ounalashkastylus tomidai,* which means "near the peninsula" in the lo-

cal Alutiik language to honor a Japanese desmo specialist named Yukimitsu Tomida.

I sent Lou and Michael several drawings of what I thought the animal might have looked like in life. We went back and forth a few times, as Lou and I debated whether desmos had large lower lips or none at all. I'm still thinking they didn't have much of a lower lip because of the size of some of the tusks I've seen in various collections. Lou's pretty adamant that the lower lip was large, though, because he thinks they would have been essential to sucking down large volumes of kelp. He and Michael think they even have the outlines of lips on one fossil. So in the end we went with the two-big-lips version for the official press release. I'm still a holdout no-lower-lip-on-my-desmos rebel.

– Ray Troll

Battle of Megiddo (this is the source of the word "Armageddon"), in which the Egyptians deployed the terrifying new technology of battle chariots for the first time. Dale's mind had moved on from the Ice Age, but it had lost none of its focus.

Mareca Guthrie and a mythical animal created by her father.

Later, back at the museum, we met Mareca Guthrie, Dale's daughter and the art curator at the Museum of the North. She showed us a remarkable unicorn skull that had been carefully constructed from parts of other more mundane animals like horses. The handiwork was so fine that it took me a long to time convince myself that what I knew was a forgery actually was a forgery. Once again, the multitalented hand of Dale Guthrie.

We spent the rest of the week working with Pat in preparation for our trip to the North Slope in the coming summer. The day before we left, Pat arranged for us to visit an experimental tunnel in the permafrost that was located in Fox, Alaska, just an hour's drive north from Fairbanks.

Between 1963 and 1969, the US Army excavated a horizontal tunnel several hundred yards into the permafrost near Fox, Alaska. The tunnel was completed as part of a program to study the behavior of permafrost. Pat Druckenmiller was the museum's holder of the key to the permafrost tunnel, and he drove us up to Fox to visit this palpable vestige of the Ice Age.

It was a typical Fox day, and temperatures were hovering around −30°F when we arrived. We post-holed through the deep snow to get to the small building at the entrance to the tunnel, where we were outfitted with hard hats and given a tunnel safety tutorial. The air in the tunnel was a different temperature from the air outside, and the result was a spectacular display of ice crystals that framed the door to the tunnel.

Once we were inside, we could see the tunnel stretching off into the distance, and our focus turned to the walls of the tunnel. Permafrost is not only frozen soil but also frozen river sediment, and this tunnel bored through sand and gravel layers that had been deposited at some point during the last Ice Age. I had a pocketknife, which I pushed into the wall of the tunnel. It went a few inches

## Drunken Soil and Drunken Trees

Permafrost is the drunken soil of the far north. It is the frozen ground that forms over repeated winters, and its depth is a measure of its history. In places in the far north of Arctic Canada, the permafrost is more than 3,000 feet deep. The area around Fairbanks is a mixture of permafrost and unfrozen soil, and you can sometimes tell by the trees that grow on it if there is permafrost below. In the spring and summer, the upper few feet of the permafrost thaw, and the resulting water is unable to drain away because of the permafrost below. That makes for swampy summer conditions. When the soil freezes again in the fall, the expansion of the ice moves elements of the soil around and literally rearranges the forest. The scrawny spruce in the permafrost around Fairbanks are often tilted at crazy angles, and they too are accused of being drunk.

and then completely stopped when it hit rock-hard ice.

This was the stuff that the Fairbanks miners wash away in search of placer gold, and this is the stuff that yielded Blue Babe. As we walked down the tunnel, we noticed the occasional bone sticking out of the wall, and in one spot we found a complete bison jawbone.

Pat Druckenmiller with a *Bison priscus* jaw in the permafrost tunnel in Fox.

It was pretty amazing to think that the tunnel had taken us back to the Mammoth Steppe when horses, bison, and mammoths were the most common herbivores, and short-faced bears and American lions were the feared predators.

That night we had dinner with Pat and Dale and their families, and we finalized our plans to hunt dinosaurs on the North Slope seven months later.

The time passed quickly, and in the middle of August, my wife, Chase, and I flew to Anchorage to meet Ray and his wife, Michelle, for the first leg of our trip to the far north. Ray had inherited a minivan from his sister Mimi and we decided to start our trip by heading south to Homer.

Anchorage is located on Cook Inlet, and it is amazing to think that this remote location in the far north Pacific was first charted by Captain James Cook. We headed down Turnagain Arm, a shallow body of water with wickedly fast tides that was named by Captain William Bligh of *Mutiny on the Bounty* fame. It is one of the better places in Alaska to see the white beluga whales.

I'd never seen a beluga so my eyes were peeled, but to no avail. At the Anchorage airport, Chase and I had seen a museum-quality diorama that showed how the Cook Inlet Natives had used upside-down trees, jammed into the tide flats at low tide, as perches to harpoon belugas at high tide. I had paused long and hard at this diorama, and the airport crowds surged past and ignored it. Strange things really were done in the land of the midnight sun.

It was pouring rain and it took forever to drive the length of the Kenai Peninsula, but we finally made it to Homer. We had come to see the Pratt Museum and understand the geology and fossils of the Kenai Peninsula. On the beach, just 2

## The Homer Tapir

In September 2017, Ray and I opened a museum exhibit of *Cruisin' the Fossil Coastline* in Anchorage at the Anchorage Museum. On opening night we met Janet Klein, a woman we would come to know as Granny J. We learned that she had been inspired by finding a mammoth ankle bone on the beach near Homer in 1991. She kept looking and found part of a tusk about ten years later. By 2010, she had organized a group of citizen scientists who patrolled the beaches of Homer in search of fossils. To date, they had found parts of nine woolly mammoths and two steppe bison.

A few months before we met her, Granny J was walking the beach with her daughter Deb, son-in-law George, and grandsons, five-year-old Kai and four-year-old Sylas. The kids were clearly being trained to look for fossils, and Kai spotted what he though was a piece of petrified wood sticking out of the sand, but when he grabbed it, he realized that it had teeth. Granny J knew right away that this was no Ice Age animal. The bone was petrified so it must have come from the Miocene bank.

They showed the jaw to Pat Druckenmiller in Fairbanks, who puzzled over it before coming to the startling conclusion that it belonged to a tapir, a reclusive tropical mammal distantly related to horses and rhinos. Today, tapirs live in the rain forests of Central and South America and those of Southeast Asia. Kai's tapir maps right onto the ever-evolving story of Wolfe and Tanai's forests and the long connection between Asia and North America.

Kai's tapir is also amazing because it is the first pre-Ice Age Cenozoic land mammal fossil from Alaska. Like the palm fronds of Kupreanof Island, it shows that Alaska has had a warm history.

Kai showed up at the exhibit the next day and Ray and I had the somewhat surreal experience of being lectured to about fossils by a little towheaded kid who had no problem at all pronouncing the words "Pleistocene" and "Miocene" and explaining what they meant.

miles west of Homer, there are 12-foot-diameter tree trunks that are exposed at low tide. These are not beluga-hunting trees, though – they are 15-million-year-old *Metasequoia* trees. There are coal seams all around Cook Inlet, and they occur in a sequence of rocks known as the Kenai Group. Where there are coal seams, there are fossil plants. And where there are fossil plants, there is a story.

In the early 1960s, the US Geological Survey scientists from California were all over Alaska. Jack Wolfe (the paleobotanist who had attended Camp Hancock as a kid) started working in Alaska in the year I was born. He collaborated with Estella Leopold, a scientist who studied fossil pollen and spores. By studying both the leaves and the pollen, Jack and Estella were able to get a very good idea of the ancient vegetation of Kenai Peninsula, and what they found was really surprising.

12 year old Devon Foster from Wasilla found this American Lion skull on an Alaskan family canoe trip in 1995.

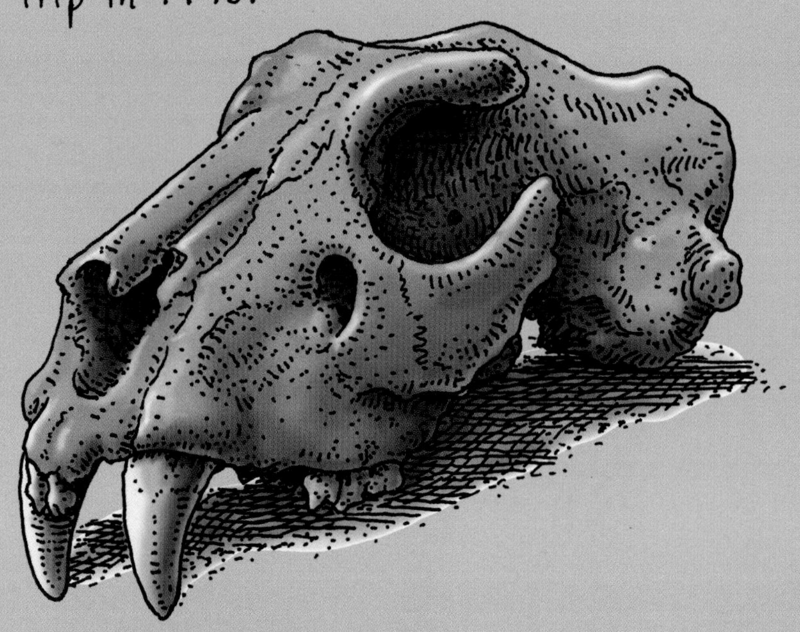

Together, they found more than forty species of broadleaf trees, including maple, liquidambar, elm, willow, poplar, alder, birch, elm, oak, and Asian wingnut. They also found the dawn redwood, *Metasequoia*. The flora was a mix of things that today live in both western North America and eastern Asia and those that live only in eastern Asia. So many Asian species were present that Jack eventually started collaborating with the Japanese paleobotanist Toshimasa Tanai to understand this forest, which would have been contemporaneous to the petrified forests of eastern Washington and fossil plants of Painted Hills of Oregon.

Tanai brought with him the knowledge of the fossil sites of east Asia and particularly the Miocene sites near Kobe and Osaka in central Japan. To these two men, the fossil record on both sides of the Pacific showed a clear pattern of connection and continuity, suggesting that western Alaska had long been a point of connection between Asia and North America. One of Tanai's Miocene sites near the small town of Toki in Gifu Prefecture was very close to where, in 1950, a middle school teacher and a high school student found the first desmostylian skeleton known from Japan. In retrospect, with high school students finding Miocene plants and desmostylians, Tanai's Gifu seems a lot like Wolfe's Oregon.

Betsy Webb, an old friend of mine from Denver, had just retired from the Pratt Museum in Homer. I had not seen her for twenty years, but she didn't miss a beat when I called her to say we were in town. She met us at the museum and showed us a very cool skeleton of the extremely rare Bering Sea beaked whale. Out back, they had just finished a long project to use horse manure to cure the bones of a gray whale and were in the process of mounting the not-too-smelly skeleton.

Betsy ushered us into the basement where we pulled drawers and delved into the fossils of the Kenai. As expected, there were many fossil leaves and shells from the Kenai Group. One unusual fossil was a football-sized rock that proved to be a split concretion containing an entire herring-sized fossil blackfish. Another drawer contained beautiful ammonites from a spot just across the Cook Inlet.

We headed back to Anchorage, stopping briefly at a wide spot in the road called Ninilchik where Ray had recently painted a large mural for a rock festival called "Salmonstock." In Anchorage, we had a cup of coffee with Tony Fiorillo from the Perot Museum of Nature and Science in Dallas. I had known Tony since we shared an academic advisor in Philadelphia, and it was good to see him. He had been researching Alaskan dinosaurs for more than a decade and had been working on the Colville River where we were headed.

Ray was getting increasingly nervous about camping on the tundra in the company of bears. Tony didn't help matters by describing grizzly bears on the Colville as either "absent or intimate." Tony had just discovered and described Alaska's first dinosaur species, naming it *Pachyrhinosaurus perotorum* after the Perot family, who supported the construction of the museum in Dallas. He had also been working in Denali National Park and finding a diversity of dinosaur footprints in the Cretaceous Cantwell Formation.

There are a lot of Cretaceous, Jurassic, and Triassic rock outcrops in Alaska and, with Tony and Pat on the job, new sites were being discovered at an amazing rate. Ray's childhood love of dinosaurs was now battling his completely reasonable adult fear of bears.

We dropped Ray's wife, Michelle, at the airport, and Ray, Chase, and I headed north. We passed the site of the recently defunct Anchorage Museum of Natural History, where we had previously seen an amazing American lion skull that had been found by twelve-year-old Devon Foster from Wasilla. An hour north of Anchorage, we passed the mouth of the Matanuska Valley. A few years earlier, Ray had helicoptered into a spot in the nearby Alaska Range where he found ammonites in concretions, and earlier in the year, he and I had visited Colgate professor David Sunderlin, who had located what we think is Alaska's northernmost palm frond at latitude 61° north.

We had fantastic, clear, warm weather, and Denali was spectacular as we drove north, finally stopping for the night in Talkeetna where Ray closed down the bar at the open mic event. The next morning, we stopped in at the Denali Visitor Center and saw the dinosaur tracks that Tony had mentioned. By evening, we were back in Fairbanks and ready to launch to the north for our Arctic dinosaur adventure.

Chase DeForest and Michelle Troll with ammonites from Cook Inlet.

## The Great Matanuska Ammonite Hunt

If I use the sharp end of the rock hammer I might be able to bury it right between the grizzly's eyes. That would slow him down. Or maybe I should just go straight for an eye. Blind him, one eye at a time: pop, pop! And if I held up this big chunk of ammonite with my left hand, I could use it like a shield to fend off his claws when he tried to rip my face off.

I was thinking all this while staring at the fresh mound of bear crap in the trail in front of me. Ever the pessimist, I was leaping to the most grim of conclusions. "Artist Ripped Asunder at Alaskan Fossil Site," blared the statewide headlines. What a way to go. But there was no bear in sight.

A couple of things were clear to me. This wasn't like any other fossil hunting trip I'd been on. There was some genuine danger in this venture. The other was that I clearly hadn't been thinking straight when I jumped out of the helicopter earlier that day.

I was in such a mad rush to find ammonites that I tossed my backpack aside and plummeted headlong toward the juicy strata below. Fossil fever had gripped me like a madman. Like a lot of things that day, it seemed like a good idea at the time. I'd just go over to the edge of the ravine, pick up a few fossils, and be back shortly for my gear. Wrong.

My friend Kirk had trained me well over the course of our five-year fossil-hunting road trip through the American West. I knew Mesozoic marine shale the moment I saw it, and we'd just been hovering over a valley absolutely chock-full of the right kind of rock. Now I was 2,000 miles away from my young mentor, and I was on my home turf of Alaska, 30 miles from the nearest road and 4,000 feet up in the Talkeetna Mountains. I could see him now in my mind's eye, chuckling at me, the woefully unprepared eager artist.

I was with a loose-knit group of fifteen folks who had chartered a helicopter to hunt ammonites in the waning days of summer. It was a spectacularly clear day, and the occasion was the twenty-fifth anniversary of the Anchorage office of GeoEngineers, a firm headed up by my buddy Scott Widness. I was crashing their company party. I had been telling Scott about the huge ammonites to be found in the Matanuska Valley

AN UNPLEASANT END TO AN OTHERWISE PERFECT AMMONITE HUNT.

for years, and he decided to celebrate this anniversary with a helicopter fossil hunt. Scott offered me a spot, and I jumped at it.

But now I was at the bottom of a steep slope staring at a pile of bear poop, brandishing my hammer like a weapon, and wondering what in the hell I was going to do. It was blazing hot, I had no water, and I was starving. I'd have to climb 1,000 feet back up through squishy calf-high tundra, and I didn't know if I could do it. The air was pretty thin up at this elevation, and I started wheezing only moments into the climb. Pretty lame right? But hey, I'm a coastal artist guy, not a mountain climber.

I could see for miles around me and nobody was in sight. It was as if the vast landscape had swallowed all of us. I wasn't even sure which slope we had landed on. Despite a few hours of searching, I had only found two fairly indeterminate and massive chunks of ammonite. The only thing to do was to push on, back up the nearest hill, zigzagging my way up through the bramble, toting my heavy treasures. Walking on the tundra is like walking on soft mattresses … only pitched at a 45-degree angle. I was one hurting old hippie after a half hour of it. Visualizing a blood-soaked grizzly standing over my broken carcass added a little motivation to my stride.

I glanced up at the ridgeline above me and saw a dark figure silhouetted against the blue sky. I stopped in my tracks, frozen. I ripped my sunglasses off for a better look and realized I was looking at a caribou. Oh, man. How cool was that? The caribou/ammonite combo package … only in Alaska.

Smiling to myself as my blood pressure dropped, I heard a voice off in the distance. It was Scott. "How ya doing?" he yelled. "Great," I bravely lied. "Got a couple of good chunks and I'm headed back up to the drop site to get some water and grub." "You don't have any water?" Scott hollered, "Man, you're nuts!" Tell me about it.

I headed toward him, back into the dark shale. I had to dig my heels in deep to keep from sliding down the hill. A cascade of rocks showered down below me with each step. Before long I had managed to work my way neatly into a dead end. I couldn't go up. I couldn't go down. Screwed again. Man, this was pathetic.

"I don't know, man, but I'd say this Matanuska Ammonite Wrestling is a young man's game. We fifty-year-old guys should just stay home!" I weakly explained to Scott when I got within 25 feet of him. He looked at me and raised an eyebrow, "Don't tell me we're going to have to sling you out of here, Ray!"

Just as I was seriously about to consider such a sad and dismal scenario, I spotted it about 10 feet above us. There, sitting in plain sight in the dark gray shale, was a glistening white 5-inch ammonite. It was perfect. Scott saw it too. I whooped out loud for the first time that day. Somewhere within me I found the energy to make a beeline for it, clambering like a 200-pound sprawling insect up through the loose shale, in a shower of flying rocks.

I held my prize in my hands and turned it this way and that. Oh, that sweet, sweet spiral, so evocative, so inspiring. In my opinion, ammonites are one of the most elegant forms in nature. I never tire of looking at them. Even though you can buy one at just about any rock shop in the world, the best fossils are the ones that come with a story. This little gem had earned some personal history with me: helicopters, grizzlies, caribou, and a healthy dose of angst.

– Ray Troll

# 13 MINIVAN TO THE POLAR FOREST

Ray, Chase, and I had driven a suburban minivan to Fairbanks, and now Ray and I were planning to drive it up the Dalton Highway, over the Brooks Range, and on to Prudhoe Bay at the edge of the Arctic Ocean. Pat Druckenmiller, Kevin May, and Florida-based Greg Erickson, a dinosaur growth-rate expert, were waiting for us back at the Museum of the North in a beefy diesel pickup truck that was clearly up to the task. They were kind, but we could tell they were a tad skeptical about our choice of vehicle. They had already packed the truck, so there was nothing to do but head off. We waved good-bye to Chase and started driving.

The Dalton Highway (also known as the Haul Road) was built as a support road for the Alaska Pipeline, a remarkable 800-mile engineering feat that connected the massive oil fields at Prudhoe Bay on the Arctic Ocean to the Valdez supertanker terminal on Prince William Sound. The pipeline

Driving the Haul Road from Fairbanks to Prudhoe Bay dreaming of the Cretaceous world of the North Slope.

was built in only three years between 1974 and 1977. The leg from Fairbanks to Prudhoe Bay is 500 miles, and the first 100 miles or so are paved, so the sunny weather and smooth pavement made us think the drive would be a piece of cake.

We passed through Fox and up into a rolling terrain covered by small tilting black spruce trees. The road was busy, and we jockeyed for space with tractor trailers, motorhomes, motorcycles, and pickups. By the time we crossed the mighty Yukon River, we thought that we had fallen behind Pat's truck so we raced to catch up. They were actually behind us and driving slowly, waiting for us to catch up with them. Eventually we reunited at the truck stop in Coldfoot, the last chance for gas or food until Prudhoe. The place was full of Japanese motorcyclists, and the walls were covered with photos of wrecked trucks. I had a patty melt and tater tots, Ray had a last glorious cheeseburger, and off we went.

The landscape grew more and more magnificent as we approached the Brooks Range, and more rocks started poking out of the tundra. Eventually we crossed the tree line and there was only tundra. Near the summit of Atigun Pass, we stopped to look for Devonian fossils but didn't have any luck.

As we drove down the other side of the pass, we entered a geological wonderland where each mountain had a different splendid exposure. I would guess the age of the rock, having no geologic map to let me know if I was right or wrong. Occasionally we would see mountain sheep on the steep slopes. The traffic had spread out, and we were feeling pretty alone. Eventually the val-

ley widened, and we passed the road to Toolik Lake, a famous University of Alaska Arctic ecology field station. And with that, the mountains were behind us and we emerged onto an endless tundra plain. The North Slope is just that – a giant, gentle ramp that rolls north from the Brooks Range to the coast of the Arctic Ocean.

Here and there we saw pairs of caribou hunters stalking their prey across the tundra, and every few miles we spotted a caribou or two. We were far north now, and the sun was reluctant to set. About 9:30 p.m., we took a little side road and rolled into a jumbled encampment of trailers, containers, plywood sheds, caribou heads, small planes, and mixed debris. We had arrived at place called Happy Valley, but it looked like the set for a Mad Max movie. It sat right at latitude 69.1° North, a couple hundred miles above the Arctic Circle.

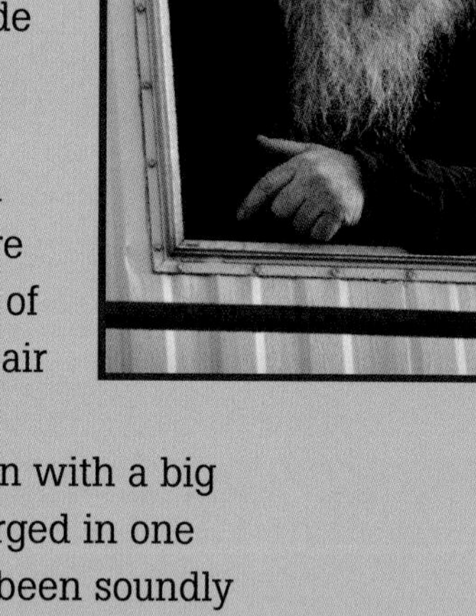

One of the containers was painted bright yellow and had a small, square room with big windows sticking out of the top of it. This, it turned out, was air traffic control for the remote airstrip. Pat honked his horn, and a giant man with a big blond beard and long gray hair emerged in one of the windows looking like he had been soundly asleep. This was Mike McCrary, the proprietor and pilot of the bush pilot business known as Seventy North. It was time to stop driving and start flying.

Upper left: Approaching Atigun Pass.

Middle left: Arriving at Happy Valley.

Lower left: Landing at Poverty Bar.

Right: Mike McCrary, our pilot.

Opposite: Patterned ground on the North Slope.

The next morning, we packed our gear into duffels, weighed them, and then loaded them into a very small plane with enormous wheels. The plane had room for the pilot, two passengers, and gear. Pat and Kevin flew first, taking a deflated Zodiac (rubber raft) and a tent with them. An hour and a half later, Mike returned to collect Greg and a mountain of gear. Another hour and half passed and Mike returned to get Ray and me and our gear.

Mike taxied and took off, and soon we were flying over the tundra. The view from the air was startlingly geometric. Repeated freezing and thawing of the active zone of the permafrost had created an incredible patterned surface to the tundra. In some places, there were circles, in others, rectangles and squares, and these shapes came in all sizes. There were places where the entire landscape looked like a giant chessboard. We had been driving past these features in the minivan, but we couldn't see them at ground level. From the air, they were magnificent.

I was crammed in the front seat next to Mike in a way that was pretty intimate. Ray and Mike soon discovered that they were the same age and had both attended high school in Kansas, so they started bonding over stories of tractors. Mike's one rule for bush flying was, "When in doubt, don't do it." We inquired about the kind of people who frequented 70 North, and Mike told us that it was mainly caribou hunters and archaeologists. Just the day before, he had picked up two hunters who had killed three caribou and a big white wolf. It hadn't even occurred to me that wolf hunting happened in a place where there was no livestock to protect, but then I remembered that we were in remote Alaska where any large mammal was fair game.

About this time, the Colville River came into view, and we could see the guys setting up tents on a big sandbar. They looked like ants. Mike circled once and put the plane down on the beach. We unloaded in a few minutes, and then he taxied and took off. For better or worse, the five of us were now at a place called Poverty Bar for the next week.

Pat had assembled and inflated the Zodiac, and Kevin was attaching the outboard motor. I walked around looking at the beach and immediately saw wolf and sandhill crane tracks. The sandy bar had patches of gravel, and some of the pebbles were pieces of coal. Here and there were beautifully preserved fossil shells. Kevin explained they had eroded out of a Pleistocene layer known as the Gubik Formation, which had also yielded bones of sea otters, seals, walruses, belugas, gray whales, and bowhead whales. About 100 yards away, the riverbank showed distinct layering, and I could see right away that some of these layers were volcanic ash beds. This

Left: The Arctic hadrosaur, *Ugrunaaluk kuukpikensis.*

Below: Ray reclining on his fossil bed of dinosaur bones.

was the Prince Creek Formation, the one that had produced the dinosaurs.

We set up our tents, and Kevin set a short wire fence around the food tent and attached the wire to a car battery. Ray asked if he could have one of these for his tent, but Kevin said there was no need since the bears always went for the food.

Once we had set up camp, we geared up and walked downstream to where the sandbar disappeared and the riverbank was right next to the river. The Colville was moving rapidly, and it was plenty cold. I made a note not to fall in.

The beach was littered with chunks of dinosaur bone. I have been digging fossils for a long time, and I have never seen such a spew of bone. Most of the pieces were ends of the leg bones of a medium-sized duck-billed dinosaur. Ray had been waiting nearly thirty years to find his first Alaskan dinosaur, and, all of a sudden, he was awash in them. All of these bones had eroded out of the bank and were lying around, out of context. We decided to see how many pieces we could collect in twenty minutes. Twenty minutes later, we had more than 300 bones. Ray, always in search of a visual pun, arranged the bones in the shape of a king-sized mattress and promptly lay down for a nap on his "fossil bed."

This site was the Liscomb Bone Bed and the reason everyone was so excited was that these were truly polar dinosaurs. The site is a rare window into the hothouse world of the ice-free Cretaceous Arctic. This is even more amazing when you take the position of North America into account. We now know that our continent has been slowly rotating in a counterclockwise direction for the last 200 million years. This means that the Colville River dinosaurs lived when this landscape was actually at a latitude that is even higher than it is today. In fact, this land lay very close to the North Pole.

We hiked back to the camp and roasted bratwurst on sticks over a campfire and washed them down with bourbon. After dinner, the guys went fishing for Arctic char in the river, and Ray started drawing dinosaurs in his sketchbook.

The next morning, the weather had changed and winds were howling down the sandbar, spinning up dust devils and whipping sand into our tents. It felt more like fieldwork in the desert than in the Arctic. The four of us geared up and hiked downstream past the bone bed and out onto a vast gravel bar that ran alongside the river for nearly a mile. Kevin had found some interesting bones on this bar on his last trip, so we fanned out about 50 yards apart from each other and slowly searched our way downstream. This is a very contemplative – almost meditative – kind of paleontology where you walk very slowly, head bowed, from rock to rock, looking at each one and trying to ascertain if it is interesting or not. Every once in a while, you bend over and pick up something for a closer inspection.

Since everything on the gravel bar had been moved by the river when it was in flood, anything we found would be out of its geologic context, but it would at least tell us what the valley had to offer. After forty minutes, I had found nothing of note. Then I noticed that Ray, Kevin, and Pat had come together. That could mean that someone found something cool, so I walked over to the group, and sure enough, Ray had found a bone – and it was a cool one. He had fist-sized vertebra of a pachycephalosaur, or bone-headed dinosaur. This was a genuinely rare fossil and one that once again told the story of the ancient connection of Asia and North America. The first pachycephalosaurs were discovered in Montana

Greg Erickson, Pat Druckenmiller, and Kevin May prospect the dangerous banks of the Colville River.

around 1860, and a little more than a century later, Polish scientists discovered them in the Gobi Desert of Mongolia. Ray's bone was just one more piece of evidence that the Bering Land Bridge was open for business in the Cretaceous.

A little while later, I found a Pleistocene whale vertebra that must have eroded out of the Gubik Formation. The presence of a marine formation so far inland was evidence that the Arctic Ocean had been more extensive during one of the interglacial periods of the Ice Age.

On the way back to camp, we stopped at the Liscomb Bone Bed and dug into the bank. This whole area is underlain by permafrost, so it's really only possible to dig in as far as it has melted during the summer. In most cases this is about a foot or two before you encounter a solid wall of ice. This partial melt creates a dangerous situation because the entire slope is destabilized by the melting, and the paleontologists digging at the bone bed risk being buried in landslides that regularly cascade off the steep slopes. Something similar to this is what had killed Liscomb himself, so we treated the slumping slope with respect and caution.

Solid ice is not a problem that dinosaur diggers typically encounter, and the frozen ground has proven to be a challenge for everyone who has come to this site. As more people heard about the site and its challenges, different ideas emerged on how to dig it. In 1987, a paleontologist named Tom Rich visited the site and nearly got buried in a landslide. He came up with the idea of digging a mine shaft straight into the hillside and mining the dinosaurs underground to avoid the danger of the landslides. He tried this approach in March 2007, and approached the site by vehicles across the frozen tundra and river and used dynamite to blast open a mine shaft. When he returned to extract the bones in August of the same year, he found that

The Arctic pachycephalosaur, *Alaskacephale gangloffi*.

the Colville River had flooded during the spring and had partially filled his mine with water, which had frozen into a solid block of ice. He was able to excavate some material underground, but it was a tremendous amount of work for a little reward. The saga of the ice mine appeared on a *NOVA* episode entitled "Dinosaurs on Ice" that aired in 2008.

We dug carefully into the bone bed and discovered that it was actually a layer of brown clay about a foot thick. There were a lot of bones in the layer, but they were isolated from each other and each bone appeared to float in the clay. The other interesting observation was that the bones all appeared to belong to young dinosaurs. Greg's specialty is using growth rings on leg bones to measure how old individual dinosaurs were

Greg Erickson standing next to a layer of white volcanic ash that filled a series of dinosaur tracks.

when they died. He estimated that most of the bones we were finding in the bone bed came from hadrosaurs that were five to six years old when they perished. Pat mentioned that the layer stretched for a long distance and was rich in fossils the whole way.

Each fossil site is like a crime scene, and paleontologists are like detectives. We began kicking around scenarios of what it would take to form a mud slurry full of disarticulated baby hadrosaurs. Tony Fiorillo had suggested that slurries of mud had flowed off of the ancient Brooks Range, entrapping and transporting the young dinosaurs. This did not ring true to me. I recalled that a number of dinosaur bone beds in Alberta had been attributed to flooding associated with coastal storm surges on a very flat coastline. And I had just been reading about pig farms in North Carolina and how hundreds of pigs drowned in a recent hurricane when sea level flooded a shallow coastal plain. I imagined how their carcasses might float around for a while before decaying and falling apart. If it worked for Alberta horned dinosaur and Carolina hogs, maybe if could work for Alaskan juvenile hadrosaurs.

We ate beef jerky for lunch and curried Spam for dinner. We drank water filtered from the muddy river and bourbon. It was becoming clear that we had come to the river with the right people but with the wrong food.

We explored upstream in the Zodiac. In places where the river sliced the tundra, we could see cross sections of the ice wedges that created the patterned ground. As we motored along the shore, we could also see that there was layer after layer of volcanic ash. Volcanic ash contains tiny crystals of zircon that formed before the volcano erupted, and these crystals contain radioactive uranium, which decays to lead over time. For this reason, each crystal is a clock, and every ash bed presents an opportunity to precisely date a rock layer. One of these layers had been analyzed and gave an age of 69.2 million years. The presence of many ash layers hinted at the possibility of creating a series of dated levels, which would put these Arctic dinosaurs into context with dinosaurs elsewhere in the world.

We stopped at a place called Pediomys Point. *Pediomys* is a type of small Cretaceous mammal, and this was the site where Berkeley's Bill Clemens discovered Alaska's first Mesozoic mammals. Sometimes what you don't find is as important as what you do find, and the team had failed to find any evidence of alligators or crocodiles. These animals are common in most Late Cretaceous ecosystems, and their absence here suggested that these extinct polar ecosystems must have had chilly winters.

When we hopped out of the boat, we immediately noted fresh bear tracks. We looked around for bear, then started looking for dinosaur. In this spot the river rushed right alongside the spalling cliff, and we were more concerned about the crumbling cliff than the potential for bears. At one point, a single piece of rock the size of marble fell off the top of the hill and hit me on the top of my head. It hurt a lot more than I expected, and that little event calibrated my caution.

Fresh brown bear tracks, not a pretty sight when you are sleeping in a tent.

As I walked along the bank, I noticed a thin layer of white volcanic ash and began to follow it down the length of the outcrop. I had begun measuring and sampling each ash with the idea of dating them. Ash beds are interesting not just because you can date them but also because they fell out of the sky and landed on the landscape. This means that they sometimes preserve aspects of the landscape that you might not otherwise see.

Along the base of the cliff, I began to notice that the ash bed was not continuous and sometimes appeared as little pods about a foot wide. Looking more closely at one of these, I noticed that it had three evenly spaced, downward-pointing lobes. The penny dropped, and I realized that I was looking at dinosaur tracks and that the volcanic ash had rained down on a landscape where dinosaurs had been walking!

Just then, I heard a yell and saw Greg tumbling down a talus slope toward the river. He landed on his feet just above the water level and looked up the hill where a puff of dust betrayed that he had just ridden a rockslide. This place was really interesting, but it wasn't very safe.

Later that afternoon, Greg and I found another ash layer and began to dig it out so we could measure and sample it. It's pretty simple work. All you have to do is create a clean vertical face so you can take an accurate thickness measurement. A falcon was screaming overhead, and we named it the Falcon Ash so we could remember it later. The ash was really wet and soggy from the melting permafrost, and the sample was more like toothpaste than rock. We didn't mind, though, because the little crystals would be preserved in either case. I grabbed a chunk of soggy ash from the bottom of the layer and turned it over. There, to my amazement, was a fossil fern.

Above: Kirk and fossil ferns found on the bottom of a layer of volcanic ash.

Right: The Arctic ceratopsian, *Pachyrhinosaurus perotorum.*

When ash falls on landscapes, it sometimes buries and preserves vegetation. Many times in my career as a paleobotanist, I have carefully inspected the bottom sides of ash layers and many times I have found beautiful fossil plants at the bottom of the layers. I call this sort of situation a "plant Pompeii." These situations are scientifically quite valuable because they preserve the vegetation as it grew on the landscape, thus presenting a real view of an ancient world.

It was really exciting to find an Arctic plant Pompeii, but I quickly realized that the toothpaste texture of the ash was going to be a problem when it came to collecting the fossils. We spent a few hours exposing more fossils and found more ferns and a variety of broadleaf herbs. This was enough to give us a snapshot of the Cretaceous meadow that was one of the habitats of our polar dinosaurs. With one lucky find, we added the third corner to a triangle that included the modern tundra, Dale Guthrie's grassy Mammoth Steppe, and our Cretaceous hothouse fern meadow.

It was a great day of discovery, and I excused myself to take a personal break. There is a common story that some of the best fossil finds come when people stop looking for fossils and go off to relieve themselves. And so it was that I was sitting there on the tundra, thinking about Cretaceous meadows, when I glanced over to my left and saw part of a mammoth tusk sticking out of the tundra. In the High Arctic, the past is never far away.

We packed up our toothpaste ferns and the outhouse tusk and hiked back to the edge of the river where we had left the boat. The clouds had come in and it was gently drizzling. Greg pulled out a little bag of plaster and mixed it with water to make a cast of the grizzly bear footprint. We were so engrossed in this task that we were startled when we heard a man's voice say, "Hi there."

We looked up and floating right next to us was a guy in a tiny inflatable kayak. We certainly didn't expect that. His name was Darrell Gardner, and he had quite a story to tell. He had been walking and paddling from the Mexican border to the Arctic Ocean, and he had found us when he was only 40 miles from his goal. I thought it was pretty ironic that Ray and I had been making the same trip for the last several years, and here was a guy that was doing the 5,700-mile-trek on foot and by kayak. It always pays to be humble. We invited him to join us for dinner in our camp.

Back in camp, we continued our daily diet of jerky, Spam, and bourbon. Darrell wisely chose to eat his own food. He was traveling so light that he only had a knife and a spork. Ray was excited to meet him so we could compare notes, but Darrell was a man of few words, and he was up and gone by the next morning.

Kevin manned the cook tent and used a satellite phone to maintain contact with the museum in Fairbanks. On the morning of August 27, he told us that Denali National Park had just reported its first fatal bear attack. A forty-nine-year-old man from San Diego was hiking not far from the visitor center where we had seen the dinosaur tracks ten days earlier. His digital camera showed that he had taken twenty-six photos of the bear over a period of seven minutes. In the photos, the bear did not seem to take notice of the man. Pat showed Ray how to use our bear gun.

At night, we had been storing the heavy pistol in the cook tent where it hung holstered from the center pole. When we ate dinner in the tent, we all sat on the ground around the pole, and when the wind blew, the gun swung lazily above our heads like the sword of Damocles. I trusted that the safety was on but personally prefer bear spray over handguns.

It started raining, but we had heavy rain gear so it didn't really matter. Ray and I hiked down to the Liscomb Bone Bed where I wanted to try an experiment. I was interested in measuring just how many dinosaur bones could be found in a certain volume of rock. We found a place where the base of the bone bed was flat and distinct and where two big vertical cracks intersected in the hill, giving me an easily measured volume of rock.

Ray enjoying the Arctic weather while excavating Alaska dinosaur bones.

We spent the day slowly picking though the mudstone with our pocket knives, extracting the little bones, one by one. It was amazing how abundant the bones were and how easily they popped out of the soft matrix. Ray was living his dream of excavating Alaskan dinosaurs. Our pile of bones grew steadily as we picked away. All of the bones belonged to juvenile hadrosaurs, and some of them showed clear marks of being

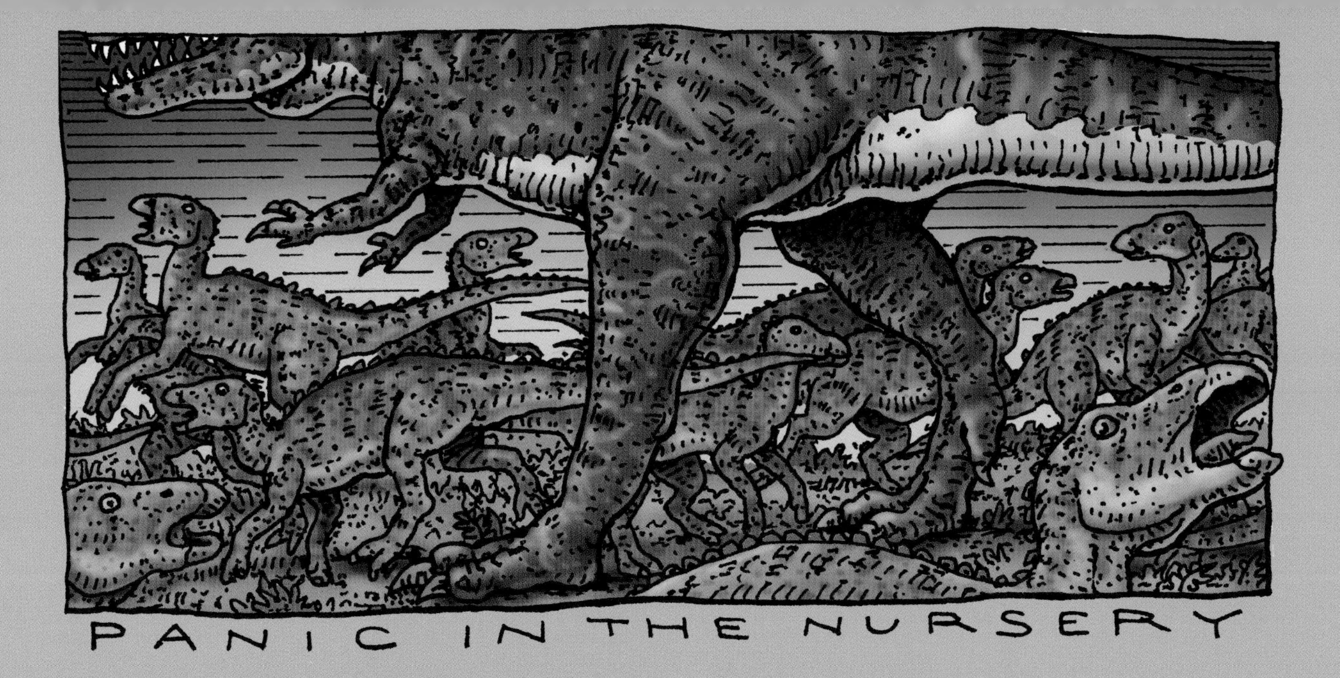


bitten by some sort of carnivorous dinosaur. After a few hours, I found a perfect serrated *Troodon* tooth. *Troodon* was a small meat-eating dinosaur, so here was direct evidence of who had been munching on the young hadrosaurs.

At the time, Pat and Greg identified the duck-billed dinosaurs as *Edmontosaurus*, a genus that is well known from the American West. Several years ago, Ray and I had visited an *Edmontosaurus* bone bed near Lusk, Wyoming, but the animals there had all been adults. During our visit, we'd wondered what had caused all of the "Dead Eds." The question was still in our minds as we picked apart the remains of this vast Arctic dinosaur nursery. Maybe this had been a rookery that has been flooded, drowning an entire generation. Had the carnivores been there before or after the babies died?

Tooth of the theropod dinosaur *Troodon*.

The dig was really fun. The rock was like butter, and the beautiful little bones popped out one after another. We were both deep in the Cretaceous Arctic. Ray kept saying that his childhood dream come true would be to find a *Tyrannosaurus* tooth in Alaska. I laughed at that and encouraged him to keep digging and he might find a *Troodon* tooth. Just then, Ray's knife clicked into something hard, and he flicked open the piece of mudstone to reveal a big carnivore tooth with tiny serrations. I grabbed it and took a hard look. I could not believe my eyes. Ray had just found his *Tyrannosaurus*. That was a moment, Ray shed a tear, and I used his eye as a scale to photograph the tooth.

Above: The moment of truth when Ray finally found a tooth of an Arctic tyrannosaur, *Nanuqsaurus hoglundi*.

In 2014, Tony Fiorillo and Dallas museum fossil preparator Ron Tykoski named the Prince Creek tyrannosaur *Nanuqsaurus hoglundi* in honor of the Inupiat people of the Colville River and Forrest Hoglund, a Dallas philanthropist. In 2015, Pat Druckenmiller and his former sudent Hirotsugu Mori named the hadrosaur as a new genus and species, *Ugrunaaluk kuukpikensis*. The name based on Inuit words means "ancient grazer of the Colville River." A fully grown *Nanuqsaurus* would have been 25 feet long and would have weighed 1,000 pounds. It would have made short work of the 5-foot-long baby *Ugrunaaluk*.

A snarling pack of Arctic *Troodon* prowl beneath the northern lights.

Ray enjoyed his bliss and started sketching. I used trigonometry to calculate the volume of the 30-centimeter-thick triangular slab of rock. Then I counted the pile of bones. In a day, we had knifed our way through 0.35 cubic meters of mudstone and had found 276 bones. That works out to a bone density of 789 dinosaur bones per cubic meter of mudstone!

That night we celebrated over a meal of leftover unrefrigerated Spam curry. Three of us got sick. On this trip, dirt slopes and dinner were definitely more dangerous than grizzly bears.

The next morning, I crawled out of my tent and noticed a fossil beluga whale vertebra that I had somehow not noticed for the whole week. Ray and I packed our bags and got ready to leave. Pat, Kevin, and Greg planned to fly out later in the day. A few minutes later we heard the plane overhead, and soon we were loaded up and flying away from Poverty Bar. This time we had a different pilot and a different flight plan.

We headed north over the Inuit village of Nuiksit and down the Colville River to the delta where it flowed into the Arctic Ocean. The patterned ground below us was too beautiful to be believed. Then we banked east and flew toward Prudhoe Bay. Soon, we were seeing drilling rigs and pumping stations, each one connected to the other by a gravel road across the top of the tundra.

The Prudhoe Bay complex stretched for miles and eventually its industrial center appeared in the distance. The oil and gas industry provides about 90 percent of Alaska's revenue, and the vast majority of it was happening right here. All of this oil would flow through the Trans-Alaska Pipeline before being loaded onto tankers in Valdez, where it would enter the world economy. Our trip began in the oil fields and Ice Age tar pits of Los Angeles, and it was ending in the oil fields and hothouse dinosaur beds of the North Slope. The irony was not lost on us. When fossil fuels are burned, carbon dioxide is released to the atmosphere, and the climate warms. Seventy million years ago, it was a carbon-dioxide-rich atmosphere that caused the Arctic to be ice-free and warm. We had just seen that world. We will likely see it again. But it was time for us to get back in our fossil-fueled minivan and drive all the way back home. The irony of that was not lost on us either.

We landed on a huge gravel strip and taxied over to a hangar. Mike McCrary was waiting for us with our minivan. Strapped to its roof was a perfect pair of caribou antlers. The minivan had made it to the polar forest and had graduated to become the Vanibou.

Ray paid for our flights by exchange, promising to design a "Happy Valley Seventy North" T-shirt for Mike. Then we walked into the Prudhoe Bay complex to get dinner. We were desperately happy to see a buffet with no sign of curried Spam.

The Anchorage paper in the café reported that 2012 was a record low year for sea ice, and that Exxon was pulling its drilling rig out of the Arctic Ocean. Mike told us that three hunters flying out of Happy Valley had just killed six bears. So the score for the week was humans: 6, bears: 1. The age of humans, also known as the Anthropocene, comes in two flavors: individual acts by people and global acts by humanity. In Prudhoe Bay, we could clearly see evidence of both.

We drove the 100 miles back to Happy Valley and had a good night's sleep in a soft bed in a plywood shed. It was good to be back in the land of food.

~Ray Troll 2014 & Grace Freeman 2014

## The Alaskan Buzz Saw Shark Comes Home

An Inupiat rock-and-roll-playing whaleboat captain called me from out of the blue while I was working in my studio in January of 2014. I was surprised, to say the least, and thought at first he was pulling my leg or that it was a wrong number, but the fellow on the other end of the line politely explained that his name was Richard Glenn and that he worked for the Arctic Slope Regional Corporation. He said he was a geologist by training and that in 1983 he'd found a *Helicoprion* whorl in Atigun Gorge a mile or so off the Haul Road north of Fairbanks when he was working on his master's thesis at Fairbanks. Within moments, I could tell that Richard knew exactly what he was talking about. The fossil was found in rock of exactly the right age, and it was evident that he was quite familiar with the complicated geology of northern Alaska.

He'd heard from Nancy Anderson at the Alaska SeaLife Center that my buzz saw shark show would be opening that summer in Seward. He told Nancy his tale about finding one in Alaska, and she thought I really needed to know this so she gave Richard my phone number. I was dumbfounded at the news and absolutely thrilled that my beloved shark had been found in my home state. We absolutely had to have the fossil in the show.

The problem was no one knew where it was. When Richard first found the strange-looking rock, he didn't know what it was. He was high on a ridge when he came across a weird-looking white coil on a black rock. It was a fairly large rock too, but he thought it was worth dragging down the mountainside to show his advisor, state geologist Charles "Gil" Mull. The two of them thought it was maybe an ammonite but unlike any they'd ever seen before. When they returned from the expedition they decided the rock should be sent to the Smithsonian for identification. A couple of weeks later they got a note back from Smithsonian curator Tom Dutro that said it was an ancient shark named *Helicoprion*. Richard mentioned the find in his thesis paper and never saw the rock again.

It had been thirty-two years and Alaska needed its shark back, but where was it? Everyone assumed it was still in the Smithsonian collection, but Dutro had passed away in 2010. I had been to the Smithsonian's vast collection rooms a few times over the years so I reached out to their vertebrate fossils collections manager, Dave Bohaska. If anyone could locate Richard's rock, it was going to be Dave.

But the trail ran cold because Dutro's office no longer existed, and his collections were in an entirely different area of the building, since he had officially worked for the USGS and not directly with the Smithsonian. I was beginning to lose hope.

Dave looked high and low and scoured case after case. We'd all but given up hope when he finally came across the fossil in an unmarked drawer one morning in March. There was much rejoicing as the email made the rounds that morning between D.C., Texas, Ketchikan, and Utqiagvik (formerly known as Barrow). Richard happened to be out in D.C. later that spring and was reunited with his long-lost treasure.

The SeaLife Center filled out all the appropriate paperwork, and the fossil was loaned back to Alaska that summer. We had a press conference in Anchorage that May to announce the rediscovery, and I finally got to meet Richard and his family, who all traveled down from Utqiagvik to see the fossil.

– Ray Troll

Richard Glenn.
(Photo courtesy of Dave Bohaska)

Big man, little trees.

Our story should end here, but it didn't. We still had to get home. We left Happy Valley at dawn and headed south. Just past Toolik Lake, we passed an interesting outcrop that we would later learn had been the source of Richard's Helicoprion tooth whorl. It was a good thing that we didn't know it at the time because we would have been forced to stop. It was snowing as we crossed Atigun Pass, and we hot-footed it to Coldfoot where we had a warm lunch. We both ordered cheeseburgers to celebrate our safe return to civilization.

Ray was driving and he would occasionally take his hands off the wheel and use them to frame an imaginary photo. I think he was trying to convince me to take a photo, but it just scared me because he was driving a Vanibou on the Haul Road with no hands. We stopped at the Yukon River and had a great slice of rhubarb pie and arrived at the Howling Dog in Fox just in time for dinner. Pat called on the satellite phone to say that fog had rolled in and they couldn't fly out. They were also running out of Spam. We had another beer and toasted them.

Our long trip up the coast was nearly over, but we were still a long way from home, so we headed for the Yukon. Our drive took us through the town of Tok, Alaska, and eventually to Delta Junction, where we stopped to have lunch with my friend Judy Olson who had married an Alaskan named Whit Hicks. We sat on their sunny porch drinking wine and talking about old times. They showed us an Ice Age bison that they had found locally, and Judy told us how Whit had recently killed a bear that was trying to kill her. Alaska seems to be a dangerous place to be a bear. Their son, Woodson, had just written an ode to paleontology, so our visit was timely. This is what he wrote:

**My dream is to be a paleontologist.**
**You get to yank the bones out of the dirt.**
**Do not try to break the bones you find.**
**Rush to a museum when you find a skeleton.**
**Everlasting pride if you find a whole dinosaur.**
**Always bring your tools with you.**
**My dream is to be a paleontologist.**

Upper left: Ray and the Vanibou at Atigun Pass.

Lower left: Whit and Judy Olsen with a *Bison priscus* horn and a mammoth tooth.

Above: Ray on the road to Tok.

We continued into Canada and drove through forests of drunken black spruce and through vast open valleys. The place reminded me of Patagonia with its huge blue lakes and howling winds. Crazy clouds skittered across the sky. Eventually, we made it to Whitehorse, a place that I had dreamt of since I read my first book about the Yukon Gold Rush.

As in Fairbanks, the gold mines near Dawson were mined hydraulically, and as in Fairbanks, it was the gold miners who found the fossils. In 1993, Sam Olynyk found a 26,000-year-old mummified horse that was touted as "Canada's best mummy." In 2013, scientists announced that the world's oldest genome was sequenced from a 735,000-year-old horse from the Yukon.

Canada's fossil laws are different from Alaska's, and the fossils from the mines belong to the province, not the miners. The province had built a lovely visitor's center at the edge of town called The Yukon Beringia Center, and they operated a fossil bone repository downtown. Just to give a sense of scale to Beringia, Whitehorse is more than 1,100 miles from the Bering Strait. The center has a sublime outdoor sculpture of a giant Ice Age beaver that was nearly as tall as Ray. Inside were skeletons of mammoths, saber-toothed cats, Jefferson's ground sloths, and a diorama with a fleshed-out short-faced bear.

Ray dropped me at the Whitehorse airport, and I headed back to Denver to prepare to move to Washington, D.C., and start my job at the Smithsonian. Ray stayed on in Whitehorse for a day and visited the bone repository where Grant Zazula curated the fossils that came out of the mines. He saw countless parts of bison, horse, and mammoth. The mammoth skulls were curiously missing their tusks.

Ray still had a long drive ahead of him, and he drove the Vanibou another 500 miles south to catch the ferry to Ketchikan from Prince Rupert. Along the way, he spent a night in Hyder, the southernmost town in Alaska. After dinner, while walking down Main Street to his hotel, he ran into a big brown bear. He didn't shoot the bear and the bear didn't eat him.

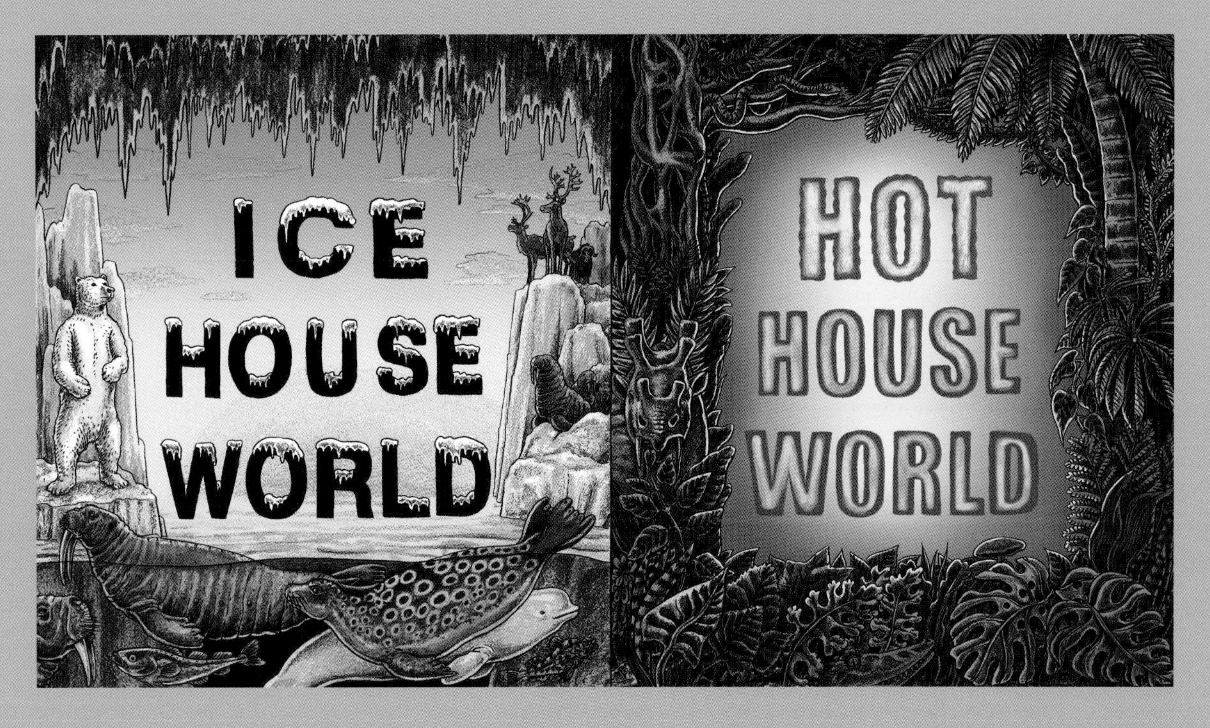

# CABO CODA

Four years later, Ray and I went to Port Townsend to see the *Western Flyer*, the old sardine fishing boat that Steinbeck and Ricketts had used to travel to the Sea of Cortez in 1940. After a strong literary start, the boat had gone back to the tough life of fishing. It fished ocean perch off of Oregon, then king crab out of Kodiak. It sank in Kodiak and was salvaged and sent to tender salmon in Ketchikan. It sank there and was taken to Anacortes where it sank a third time. The third time was not a charm, and it sank for the fourth time in La Conner. This time, someone who understood its history and meaning bought the boat. John Gregg is a mining technician and designer of remotely operated vehicles used in marine salvage and exploration. He loves fossils and history. He floated the boat and brought it to the shipyard in Port Townsend, where the restoration began. Given its history, the *Western Flyer* was in pretty good shape, a testimony to its builders and the solid wood they used. Gregg's plan was to restore the boat and use it to teach science and marine ecology in small towns up and down the West Coast.

We had not met John, but Ray was with me in Washington, D.C., when his brother, Andy, dropped by the Smithsonian. Andy Gregg told us all about John's plan for the boat and showed us a drawing of the remotely operated vehicle that he had designed to be used on the *Western Flyer*. To our amazement, the device was shaped like a large pink-striped ammonite. Ray had just finished designing the cover for this book and, as you can see, sitting on the stern of the *Nakwasina* is a giant pink-striped ammonite.

After visiting the *Western Flyer*, my wife, Chase, and I went on a cruise to the Sea of Cortez with the National Geographic Society's Committee for Research and Exploration. We were joined on the ship by a variety of scientists. Jorge Velez-Juarbe had just taken over

Larry Barnes's job as the curator of fossil marine mammals at the NHM in Los Angeles, and he was excavating sites on the Baja peninsula that were producing early sea cows and desmostylians. Susan Shillinglaw, the head of the John Steinbeck Center in Salinas came along with her husband, a Ricketts historian and squid biologist, William Gilly. In 2004, Gilly had filled a boat full of scientists and retraced the path of the *Western Flyer* in the Sea of Cortez, so he knew the spots where Ricketts had made his collections. On the last evening of our cruise, we went to Punta Lobos and wandered around the same tide pools that Ricketts had collected from seventy-five years earlier.

We never do all the things that we want in life, but we try. We are products of where we come from, passions we gathered as children, and the impressive imprints of our mentors and heroes. The world has an amazing story that is emotional at its core. We chose to tell this story as a collaboration of art and science because, for us, the beauty of things is derived from their shape, history, and meaning.

At its heart, the *The Log from the Sea of Cortez* was a story about an unfinished exploration and an ongoing friendship. So it is with our story. Steinbeck humanized science through Ricketts at a time when science and art were seen as far apart. Our intent is to use art, narrative, humor, fossils, and museums to integrate the deep time history of the West Coast and open its incredible story to people who have previously only known or loved parts of the story.

Upper right: The *Western Flyer* up on blocks in Port Townsend. Right: The tide pools at Punta Lobos on Isla Espíritu Santo in the Sea of Cortez.

A family of Alaskan desmos

# ACKNOWLEDGMENTS

Ray and Kirk thank their lovely wives Michelle Troll and Chase Deforest, who allowed a remarkable amount of unsupervised travel along the West Coast between 2008 and 2017. Funding for this travel was provided by a joint fellowship to Ray and Kirk by the Guggenheim Foundation in 2011. The text was improved and fact checked by Bobby Boessenecker, Will Clyde, and Mark Goodwin. Kirk thanks his mom Katie Jo, dad Dick, and sister Kirsten for making so many unscheduled stops at roadside outcrops between 1964 and 1978 and for being so supportive of a fairly massive in-house rock collection. Ray had help from Grace Freeman who did much of the digital coloring of his pen and ink drawings. Memo Juaregui did the color on the California map and Terry Pyles colorized the Washington State map. Frank and Jane Boyden and the Sitka Center for Art and Ecology provided two writing and drawing retreats as well as some amazing smoked black cod.

We thank Rebecca McEwen, Alison Auch, and Melanie Roth, along with Bob Baron, at Fulcrum Publishing for their great work and paleontological patience.

To gather the stories for this book, we traveled together on a dozen trips between 2008 and 2017. These trips varied in length from a weekend to three weeks. During that time, we both also made other trips alone or with other fossil friends. All told, we each spent more than 300 days exploring the West Coast for this project. Our goal was to see every fossil-bearing museum on the coast and we came pretty close to achieving it. Our journey has taken long enough that museums have opened or closed and beloved people have passed away. While this book aspired to be a snapshot, it turned into a decade.

There were a lot people who helped us access the wonderful world of West Coast fossils and we thank them all. These people told us stories, fed and watered us, put us up for the night, shared secret locality information, showed us their collections, loaned us boats, interpreted geology, referred us to their colleagues, and greatly deepened our understanding of this amazing shore. We are eternally grateful for their knowledge and generosity, and apologize to those who we inadvertently omitted. We list all of these people alphabetically by the state where we encountered them or their fossils:

## CALIFORNIA AND BAJA CALIFORNIA

Cecelia Auimopas, Josh Ballze, Larry Barnes, Tony Barnosky, Dennis Bartels, Lori Bettison-Varga, Bobby and Sarah Boessenecker, Sam Bowring, Ron Cauble, Jim Chatters, Luis Chiappe, Bill Clemens, David Cohen, Shelley Cox, Russ and Christa Crane, Lulis Cuevas, Ted Daeschler, Tom Deméré, Sara diAngelis, Bob Dundas, Mike Edward, Diane Erwin, Andy Farke, Harry Filkorn, Lori Fogarty, William Gilly, Mark Goodwin, Judy

Gradwohl, John and Andy Gregg, Lindsey Groves, Mick Hager, Ashley Hall, John Harris, Jim Hekkers, Dick Hilton, Craig and Annie Holdren, Pat Holroyd, David Hopkins, Carrie Howard, Karl Hutterer, Martin Jefferson, Ryan Jefferson, Warren Johns, Tim King, Niranjala Kottachchi, Tom Lindgren, Jere Lipps, Don Lofgren, John Long, Cindy Looy, Milton Love, Jane Lubchenco, Richard Lynas, Joan Marason, Charles Marshall, Sam McLeod, Sarah Messnick, Ed Miller, Sarah Minor, Heather Moffatt, Paul Murphey, Helen Nuckolls, Randy Olsen, Kevin Padian, Adele Panofsky, Darryl Pettit, Jane Pisano, Don Prothero, Farah Rahbar, Charles "Chuck" Repenning, Karin Rice, Meredith Rivin, Scott, Toni, and Jade Sampson, Gabe Santos, Judy Scotchmoor, Eric Scott, Eugenie Scott, Susan Shillinglaw, Doug Shore, Dave Smith, Kathleen Springer, Mary Stecheson, J.D. and Margie Stewart, Kirk Stoddard, Brian Swartz, Gary Takeuchi, Bruce, Robin and Theo Tiffney, Trevor Valle, Jorge Velez-Juarbe, Xiaoming Wang, Joe Watkins, Regina Wetzer, Lisa White, William T. Wiley, Joe Willey, Jack Wolfe, and Mike Wracker.

PALEONTOLOGISTS EARN
OIL MONEY PERFECTING
PREHISTORIC HISTORY
ALWAYS
SHE
Ray Troll 2014
PALEOCENE, EOCENE
OLIGOCENE, MIOCENE,
PLIOCENE, PLEISTOCENE,
HOLOCENE,
ANTHROPOCENE

## OREGON

Amy Atwater, Dave Bohaska, Tom Bones, Frank and Jane Boyden, Jennifer Cabin, David Craig and Kendra Mingo, Aaron Currier, Ed Davis, Regan Dunn, Nick Famoso, McKenzie Figuracion, Ted and Skyler Fremd, Kent and Lucy Gibson, Bill Hanshumaker, Jim and Jo Hockenhull, Samantha Hopkins, Alex Krupkin, Liz Lovelock, Steve Manchester, Herb Meyer, Jeff Myers, John Orcutt, Bill Orr, Nick Pyenson, Clayton Ray, Greg Retallack, Bob Rose, Josh X. Samuels, Silas Stardance, Kelsey Stilson, Bill Sullivan, Bruce Thiel, and Ed Vines.

## WASHINGTON

Brian Atwater, Ross Berglund, Ian, Jennifer, and Gavia Boyden, Bill Brandt, Cathy and Clementine Brown, Bill Buchanan, Betsu Carlson, Chris Chase, Bill Curtsinger, Jane Cushing, Tad and Rick Dillhoff, Ron Eng, Caroline Gibson, Jim Goedert, Tom, Caroline, and Beatrice Grauman, Harold Hanson, Charlie, Lucy, Harper, and Lark Hanson, Clare Manis Hatler, Estella Leopold, Stan Mal-

lory, David Montgomery, George Mustoe, Jack Nesbit, Liz Nesbit, James Orr, Brandon Peecook, Phil Peterson, Ted Pietsch, Kitty Reed, Bill Reid, Bill and Dee Rose, Christian Sidor, David Starr, Julie Stein, Mike and Jan Sternberg, Caroline Stromberg, Kathy Troost, Peter Ward, Wes Wehr, Andrew Whiteman, Greg Wilson, Bev Witte, and Nick Zentner.

## BRITISH COLUMBIA AND THE YUKON

Bruce Archibald, Graham and Tina Beard, Deborah Griffiths, Trish Guiguet, Eric Tamm, Heather Trask, Mike Trask, Pat Trask, John Valliant, Jan Vriesen, and Grant Zazula.

## ALASKA

Ketch Bachelor, Jim Baichtal, Bob Banghart, Robert Blodgett, Angela Coleman, Laura Carsten Conner, Diane Converse, Aron Crowell, Julie Decker, Pat, Lisa, and Maggie Druckenmiller, Tom Dutro, Greg Erickson, Bud Fay, Tony Fiorillo, Bill Fitzhugh, Pieter Folkens, Sarah Fowell, John, Jan and Finn Fraley, Roland Gangloff, Darrell Gardner, Richard Glenn, Dale and MaryLee Guthrie, Mareca Guthrie, Barth and Jackie and Hannah Hamberg, Bill Hanson and Kate Troll, Pete Hanson, Mary-Alice Henry, Whit and Woodson Hicks, Anna Hoover, Nadia Jackinsky-Sethi, Nathan Jackson, Lou Jacobs, Sue Karl, Deb Klein, Janet Klein, Igor Krupnik, Piers Leigh, Seth Maiers, Kevin May, Corky McCorkle, Mike McCrary, Hans Nelson, Mary Nerini, Judy Olsen, Peter Oxley, Michael Polcyn, Chip Porter, Eugene Primaky, Miyun Reid, George, Kai, and Sylas Reising, Nancy Ricketts, AlexAnna Salmon, Forrest Sheperd, Linda Slaght, Connie Soja, Gary Staab, Mat Stimpson, John and Jan Straley, David Sunderlin, Gary Thompson, Dave Tomeo, Meade Treadwell, Tim Troll, Donald Varnell, Betsy and Davis Webb, and Scott Widness.

Created in collaboration with shapeoflife.org

Special thanks to Grace Freeman, who colorized many of Ray's works of art in this book.
To see more of Grace's work, visit gracemakesart.com.

# INDEX



Heteromorph ammonite.

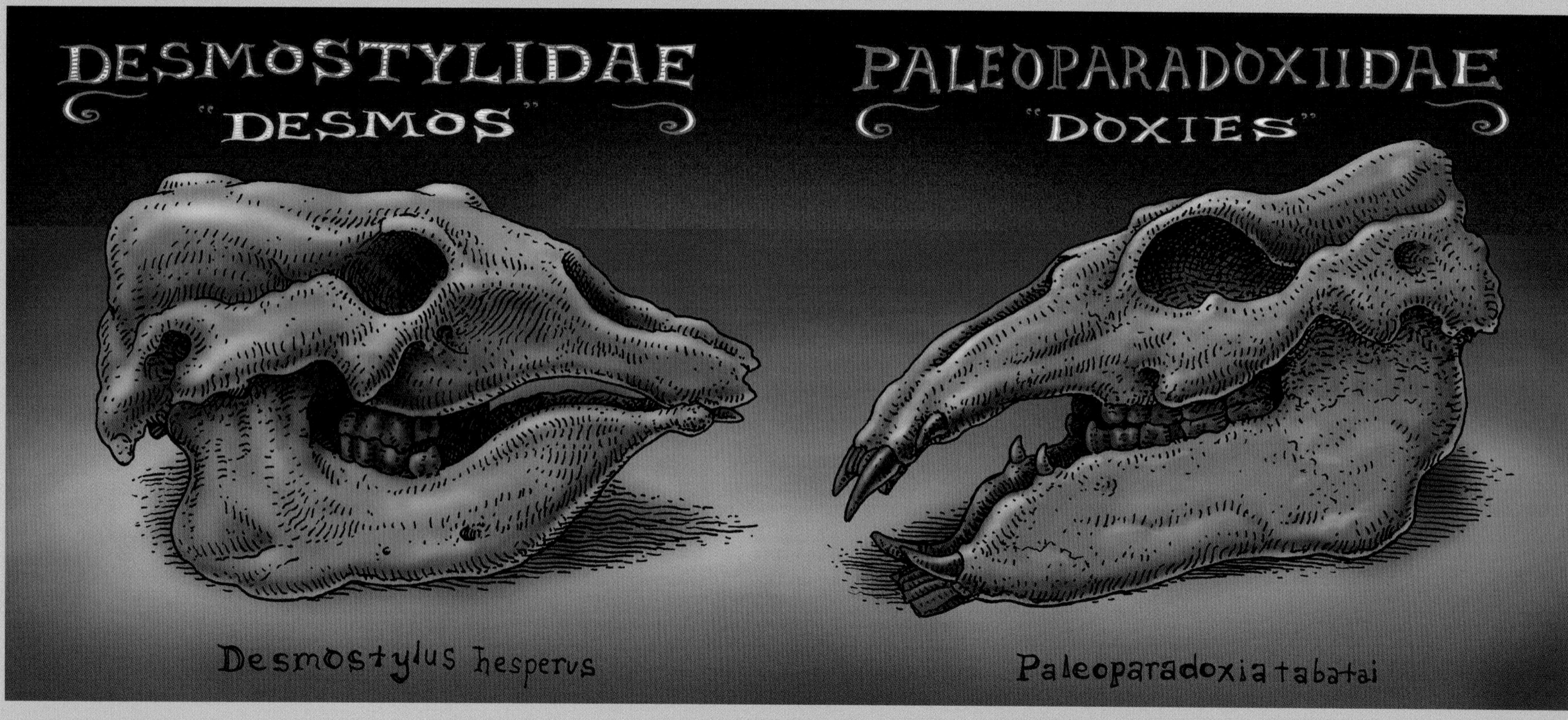
DESMOSTYLIDAE
"DESMOS"
PALEOPARADOXIIDAE
"DOXIES"
Desmostylus hesperus
Paleoparadoxia tabatai

A thalattosaur.

GORGING ON A FIELD FULL OF DEAD EDS

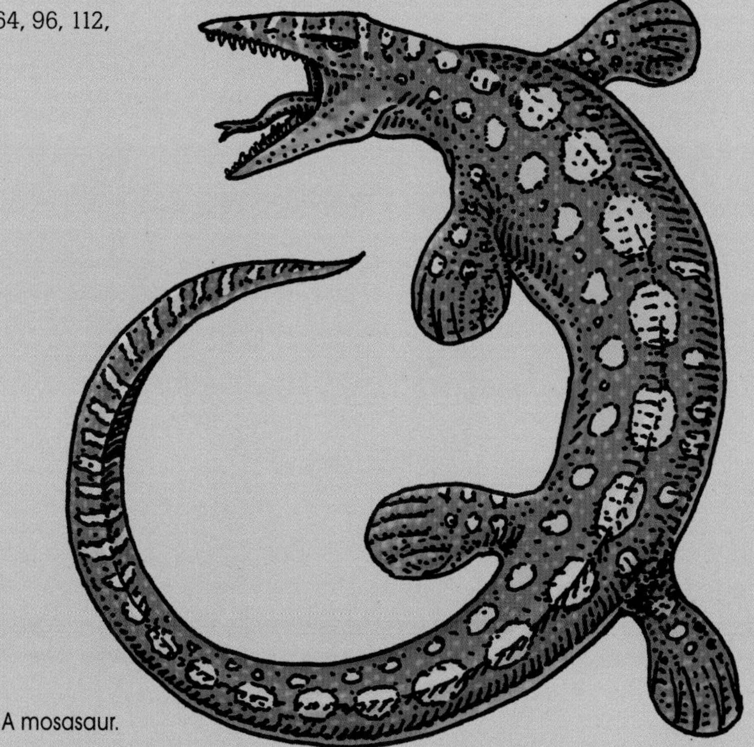

A mosasaur.

A woolly mammoth.

AMERICAN LION or a MEGA-JAG?

Helicoprion AKA the buzz saw shark.

A TYRANNOSAUR AT MY DOOR

A *Ugrunaaluk*.

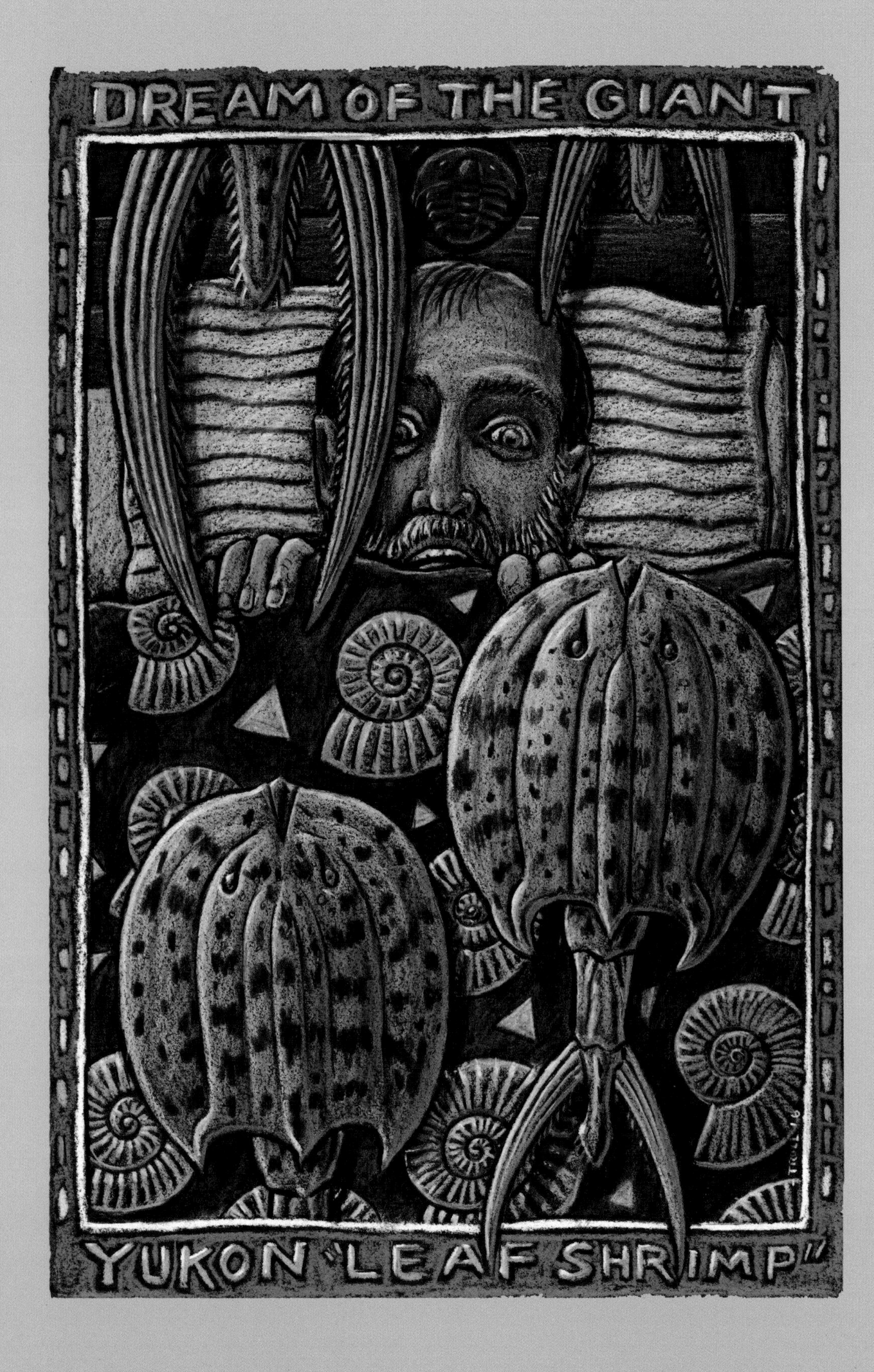




On the right: *Nanuqsaurus hoglundi*

## RULES OF THE ROAD

Finding fossils is fun, but it is important that you follow the rules if you want to collect them. For private property, always get permission from the landowner before you start to look for fossils. For state land (this includes many roadcuts, parks, and beaches), check with the state archaeologist (sadly, most states don't have state paleontologists) before you go, since rules vary from state to state. On Bureau of Land Management (BLM) land, you can collect reasonable amounts of plant and invertebrate fossils for personal noncommercial use, but you can't collect bones, teeth, or skeletons without a professional permit. Check in at a local BLM office for additional rules and recommendations. The US Forest Service has similar regulations, so check with local ranger stations to get specific details. Collecting is forbidden in national parks.

Some museums and fossil clubs organize fossil hunting trips. This is a good way to learn the rules and get your questions answered. But no matter where you are or who you are with, if you find a really cool fossil, contact a paleontologist.

**Check out the following West Coast museums, where you can see fossils and their living descendants.**

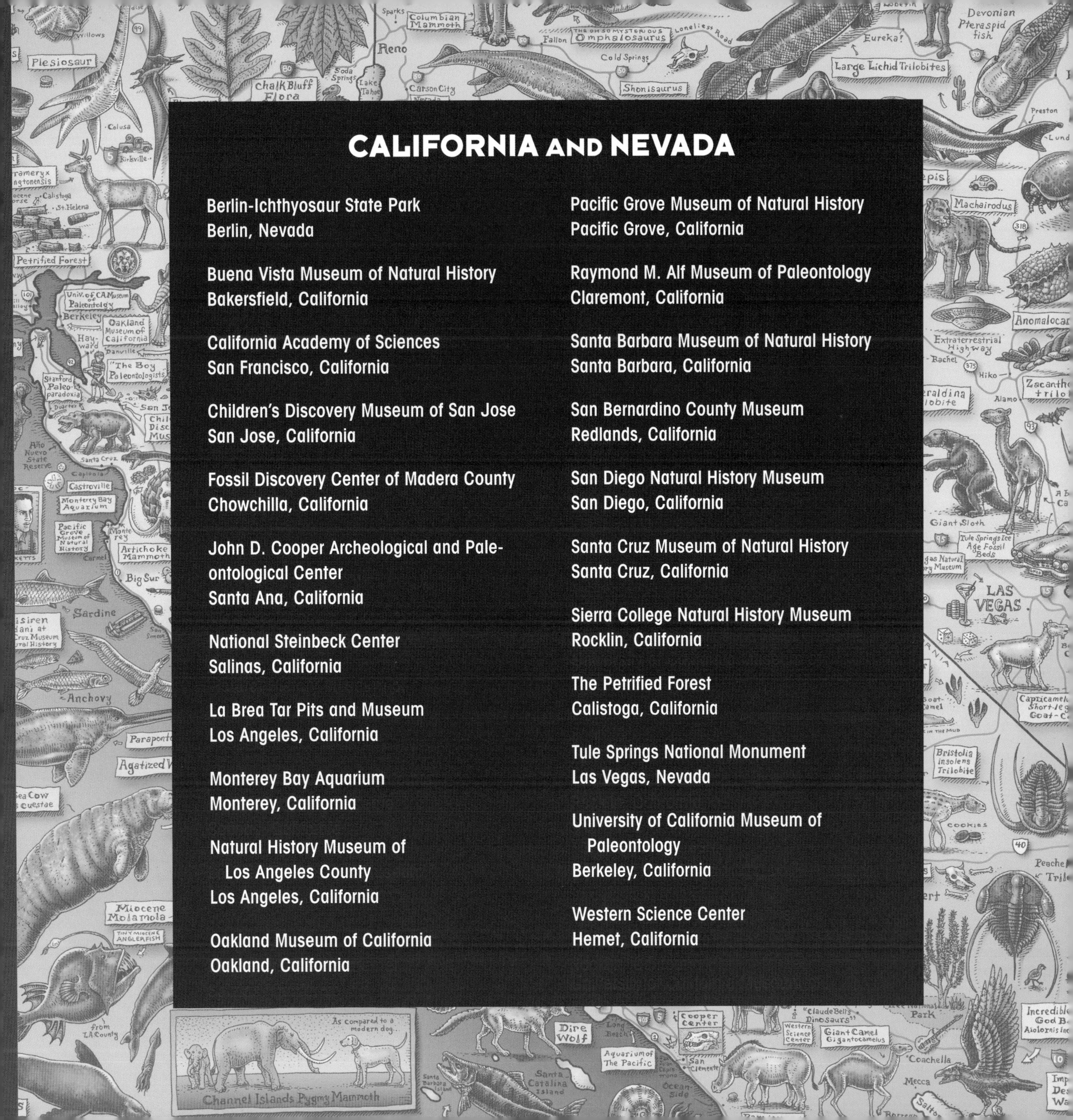


## CALIFORNIA AND NEVADA

Berlin-Ichthyosaur State Park
Berlin, Nevada

Buena Vista Museum of Natural History
Bakersfield, California

California Academy of Sciences
San Francisco, California

Children's Discovery Museum of San Jose
San Jose, California

Fossil Discovery Center of Madera County
Chowchilla, California

John D. Cooper Archeological and Paleontological Center
Santa Ana, California

National Steinbeck Center
Salinas, California

La Brea Tar Pits and Museum
Los Angeles, California

Monterey Bay Aquarium
Monterey, California

Natural History Museum of Los Angeles County
Los Angeles, California

Oakland Museum of California
Oakland, California

Pacific Grove Museum of Natural History
Pacific Grove, California

Raymond M. Alf Museum of Paleontology
Claremont, California

Santa Barbara Museum of Natural History
Santa Barbara, California

San Bernardino County Museum
Redlands, California

San Diego Natural History Museum
San Diego, California

Santa Cruz Museum of Natural History
Santa Cruz, California

Sierra College Natural History Museum
Rocklin, California

The Petrified Forest
Calistoga, California

Tule Springs National Monument
Las Vegas, Nevada

University of California Museum of Paleontology
Berkeley, California

Western Science Center
Hemet, California

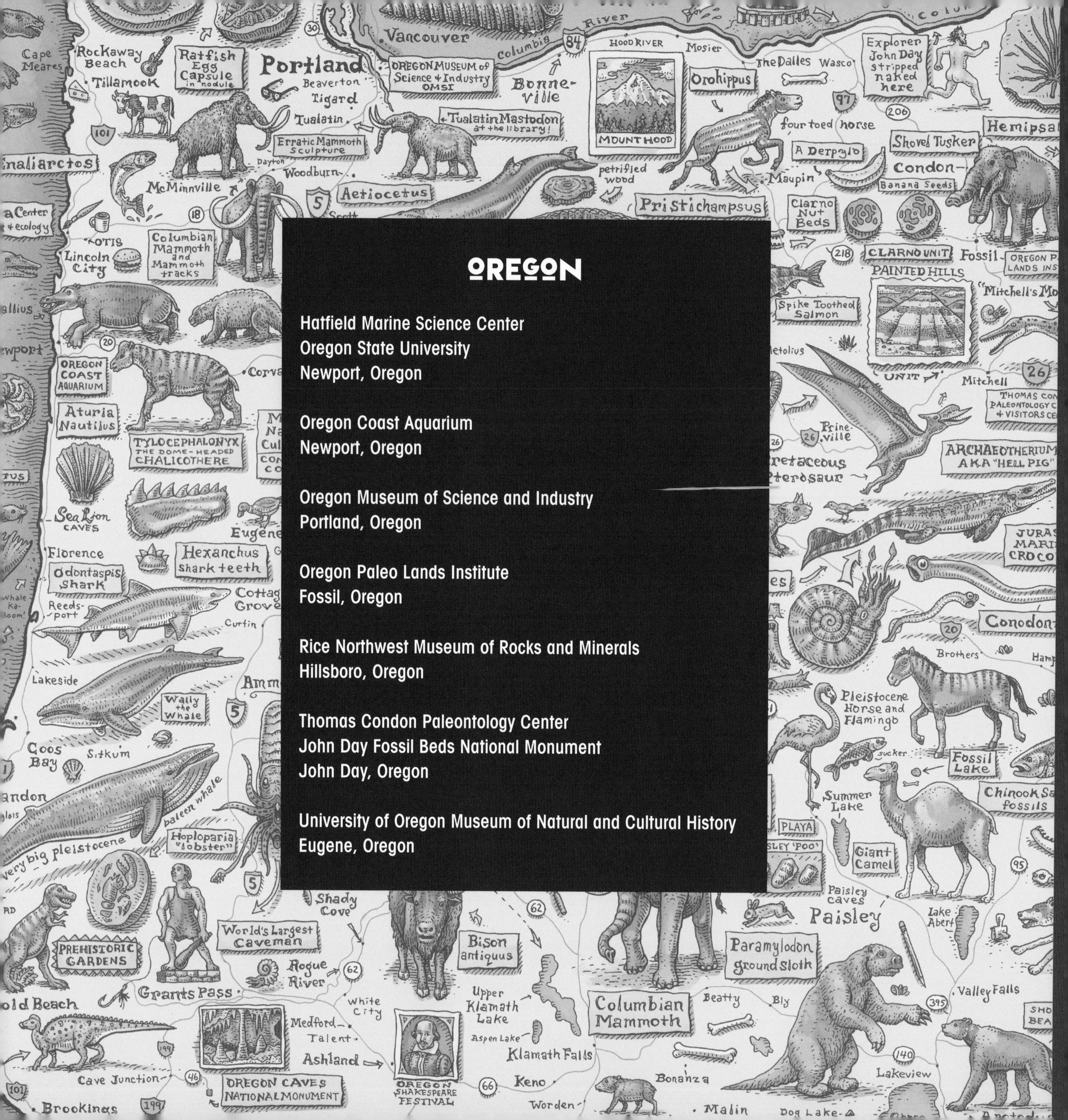


# OREGON

Hatfield Marine Science Center
Oregon State University
Newport, Oregon

Oregon Coast Aquarium
Newport, Oregon

Oregon Museum of Science and Industry
Portland, Oregon

Oregon Paleo Lands Institute
Fossil, Oregon

Rice Northwest Museum of Rocks and Minerals
Hillsboro, Oregon

Thomas Condon Paleontology Center
John Day Fossil Beds National Monument
John Day, Oregon

University of Oregon Museum of Natural and Cultural History
Eugene, Oregon

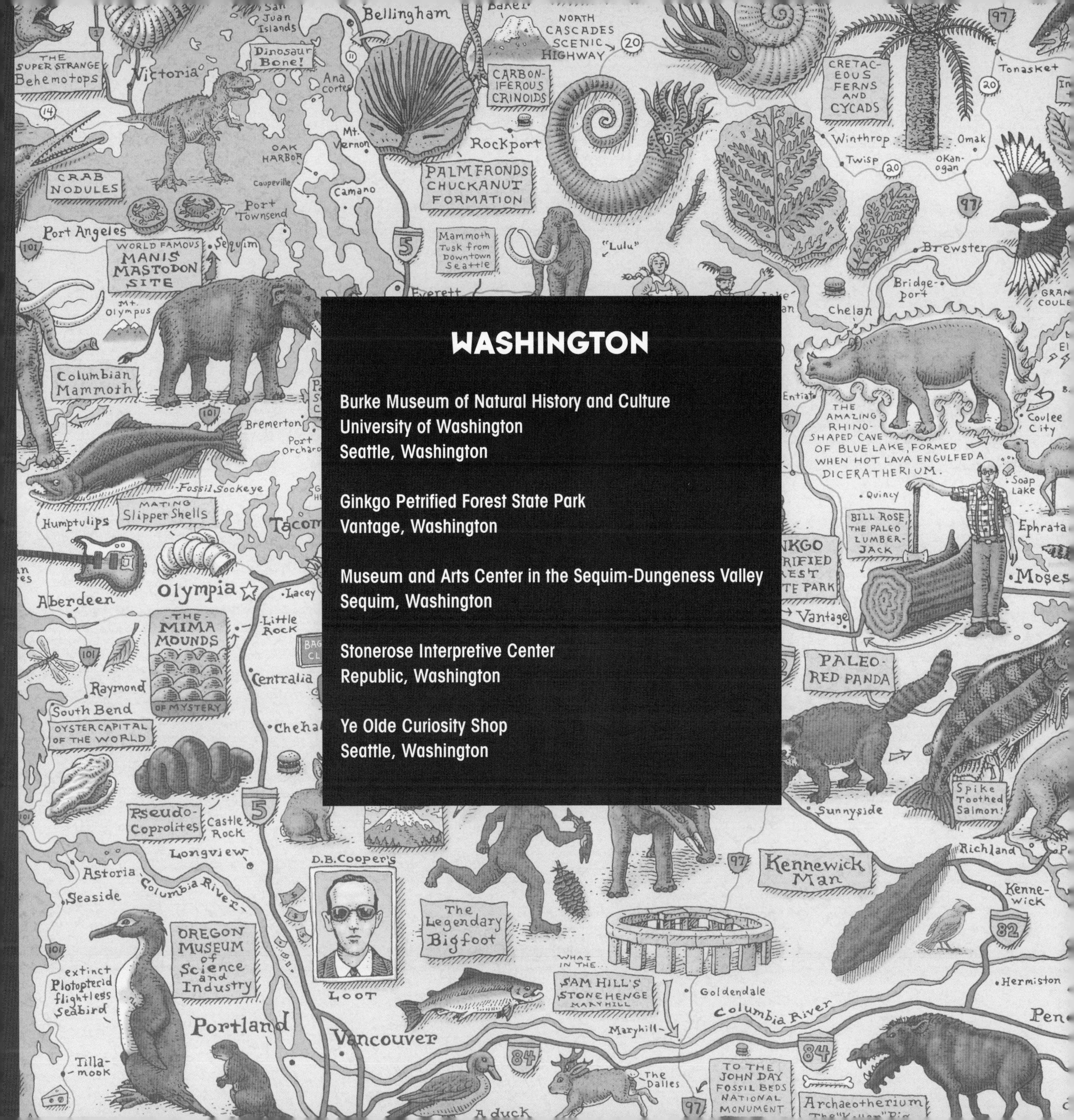


## WASHINGTON

Burke Museum of Natural History and Culture
University of Washington
Seattle, Washington

Ginkgo Petrified Forest State Park
Vantage, Washington

Museum and Arts Center in the Sequim-Dungeness Valley
Sequim, Washington

Stonerose Interpretive Center
Republic, Washington

Ye Olde Curiosity Shop
Seattle, Washington

# BRITISH COLUMBIA AND ALBERTA

Beaty Biodiversity Museum
University of British Columbia
Vancouver, British Columbia

Burgess Shale Geoscience Foundation
Field, British Columbia

Courtenay and District Museum and Paleontological Centre
Courtenay, British Columbia

Pacific Museum of Earth
Vancouver, British Columbia

Paleontology Museum
University of Alberta
Edmonton, Alberta

Philip J. Currie Dinosaur Museum
Wembley, Alberta

Royal British Columbia Museum
Victoria, British Columbia

Qualicum Beach Museum
Qualicum Beach, British Columbia

Royal Tyrrell Museum of Palaeontology
Drumheller, Alberta

Science World
Vancouver, British Columbia

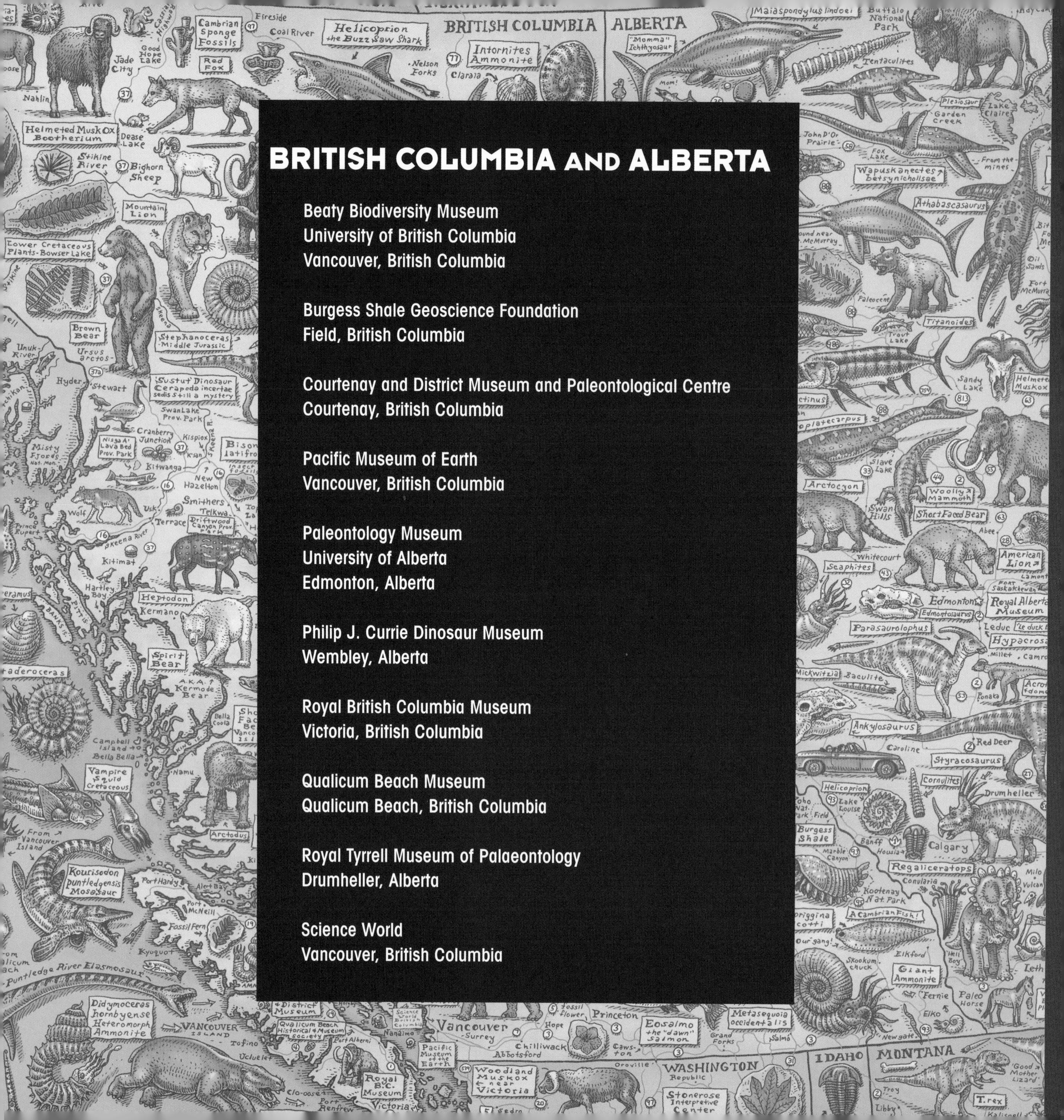

## ALASKA AND YUKON

Alaska State Museum
Juneau, Alaska

Alaska SeaLife Center
Seward, Alaska

Anchorage Museum at Rasmuson Center
Anchorage, Alaska

Museum of the Aleutians
Unalaska, Alaska

Pratt Museum
Homer, Alaska

Sheldon Jackson Museum
Sitka, Alaska

Sitka Sound Science Center
Sitka, Alaska

Tongass Historical Museum
Ketchikan, Alaska

University of Alaska Museum of the North
Fairbanks, Alaska

Yukon Beringia Interpretive Centre
Whitehorse, Yukon

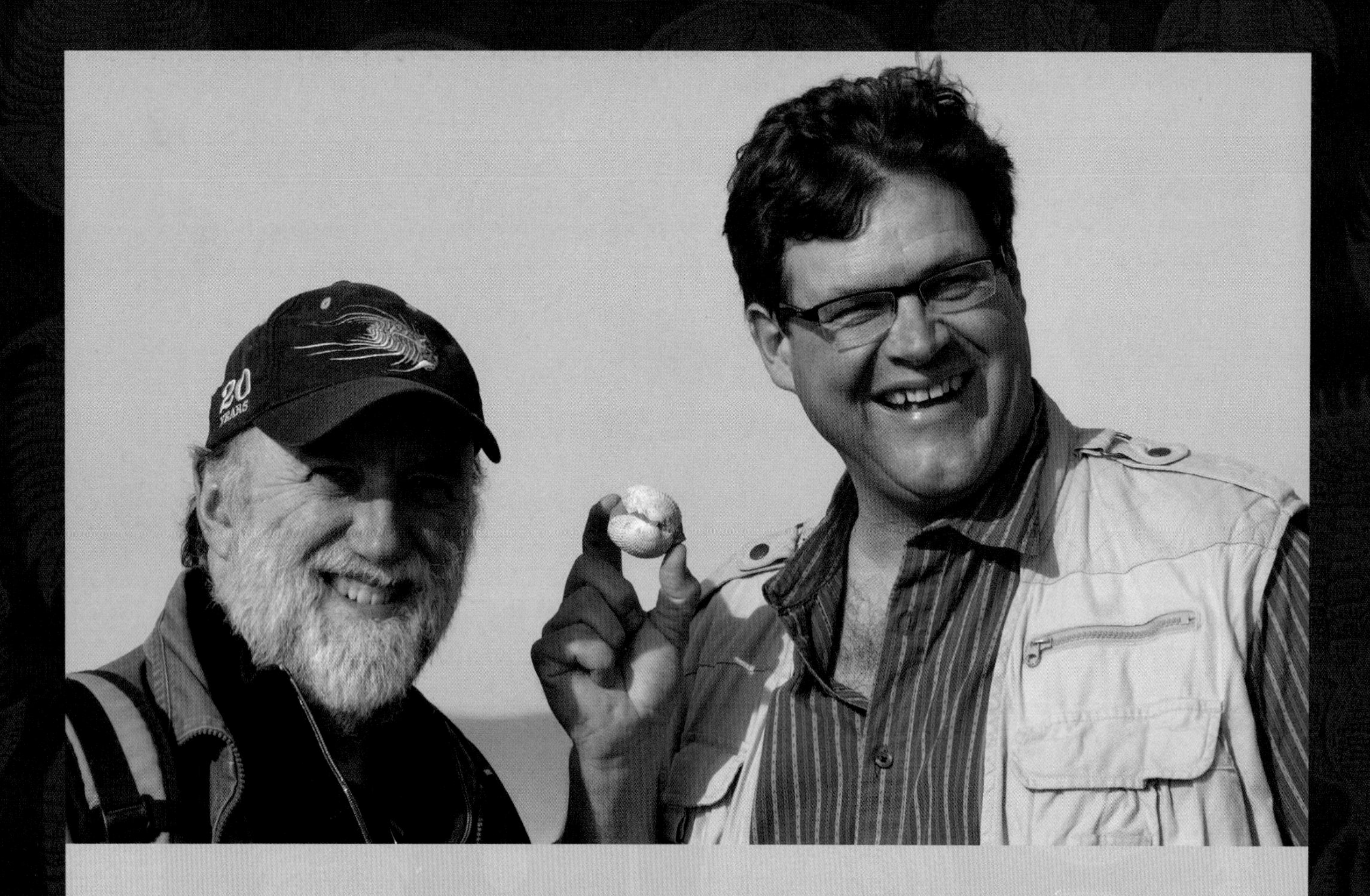

## PALEONTOLOGIST KIRK R. JOHNSON

is a paleobotanist and the Sant Director of the Smithsonian's National Museum of Natural History in Washington, D.C. With a research focus on fossil plants, ancient climates, and the K-T boundary, he strives to make his science accessible. Kirk is the host of two recent PBS series, *Making North America* and *The Great Yellowstone Thaw*. He has written ten books including *Prehistoric Journey*, *Cruisin' The Fossil Freeway*, *Digging Snowmastodon*, *Ancient Denvers*, and *Ancient Wyoming*.

## ARTIST RAY TROLL

is best known for his twisted yet accurate fish-oriented imagery. Ray has illustrated ten books, including *Cruisin' the Fossil Freeway*, *Sharkabet*, *Rapture of the Deep*, and *Planet Ocean*. His science-infused art has appeared on more than two million T-shirts. He and his wife, Michelle, own and operate the Soho Coho Gallery in Ketchikan, Alaska.

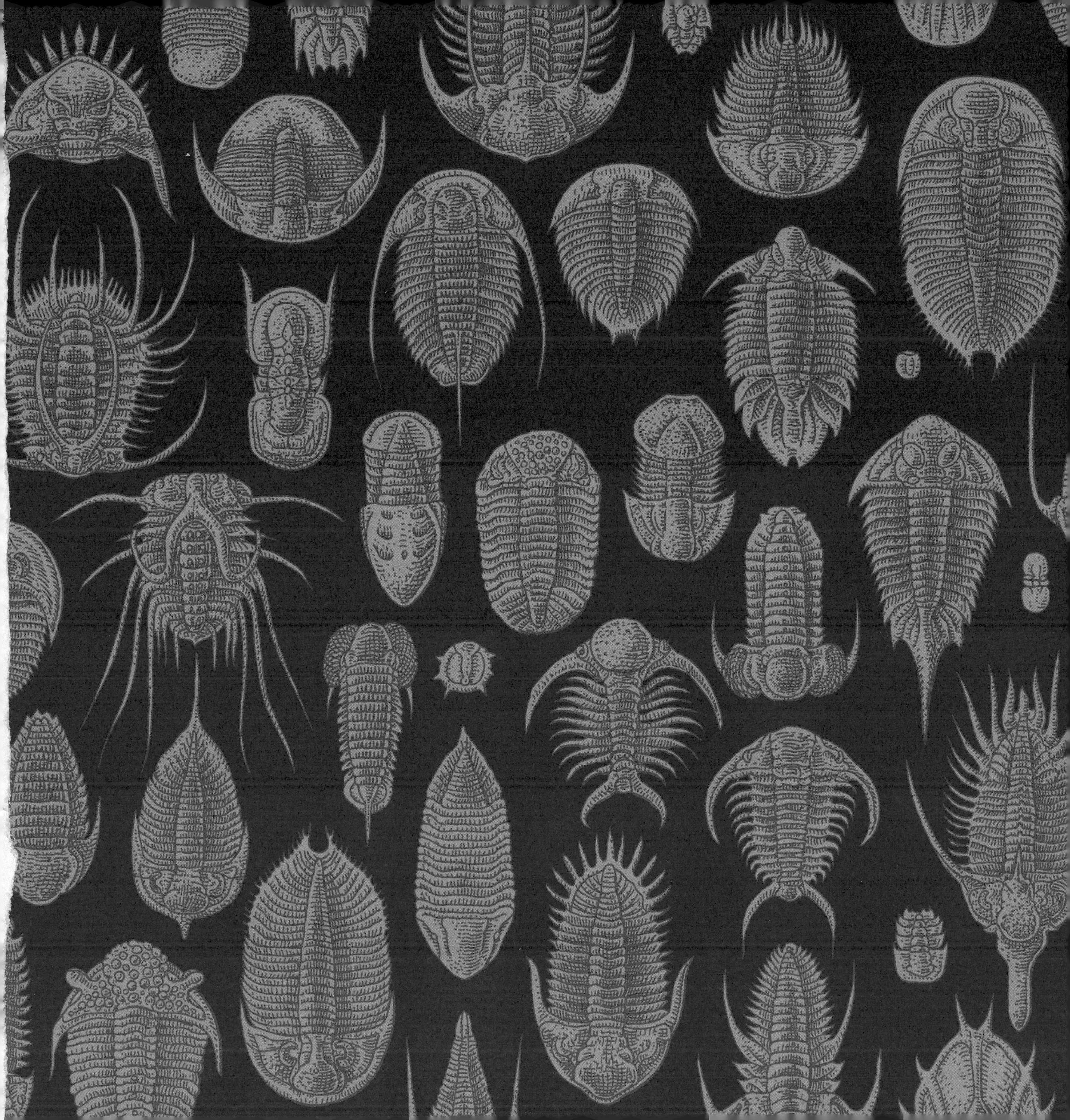

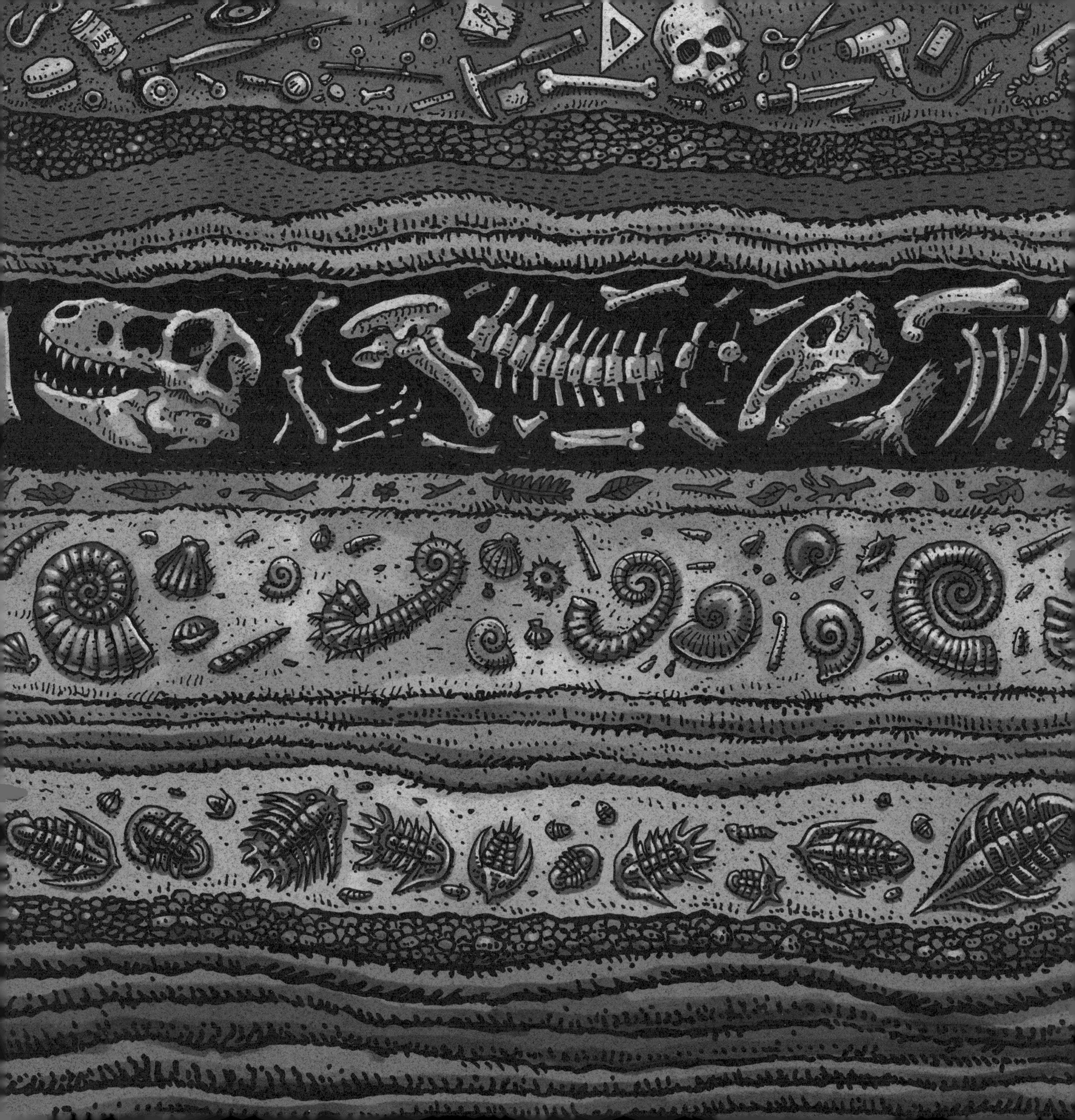